DANIEL GILBERT was born in 1957 and is the Harvard College Professor of Psychology at Harvard University in Cambridge, Massachusetts. He has won numerous awards for his teaching and research, including the Phi Beat Kappa Teaching Prize, a Guggenheim Fellowship, and the American Psychological Association's Distinguished Scientific Award for an Early Career Contribution to Psychology. His research has been featured in the *New York Times Magazine, Forbes, Money, The New Yorker, Scientific American, Oprah Magazine, Psychology Today*, and more. His short stories have appeared in *Amazing Stories* and *Isaac Asimov's Science Fiction Magazine*, as well as other magazines and anthologies.

Visit www.AuthorTracker.co.uk for exclusive information on your favourite HarperCollins authors.

From the reviews of *Stumbling on Happiness*:

'The joy of this book lies in the reading ... there are dozens of books that contain the science, but this one pulls it all together in the neatest, wittiest way possible. If you didn't laugh at the parade of human folly laid out in this book, you'd have to cry. In fact, if you are like this reviewer you will giggle, splutter, guffaw and then possibly fall off your chair making weird wheezing noises'   *Daily Mail*

'A witty, insightful and superbly entertaining trek through the foibles of human imagination'   *New Scientist*

'A delight to read. Gilbert is charming and funny and has a rare gift for making very complicated ideas come alive. He walks us through a series of fascinating – and in some ways troubling – facts about the way our minds work. This is a psychological detective story about one of the great mysteries of our lives. If you have even the slightest curiosity about

the human condition, you ought to read it. Trust me'
MALCOLM GLADWELL, author of *The Tipping Point*

'A cerebral, intelligent and extremely entertaining account of our lifetime quest for deep satisfaction. He eloquently combines philosophy and science to unravel the deep mystery of our baseline emotional state . . . he does for psychology what Bill Bryson did for evolution'
Scotsman

'*Stumbling on Happiness* is an absolutely fantastic book that will shatter your most deeply held convictions about how your own mind works. Ceaselessly entertaining, Gilbert is the perfect guide to some of the most interesting psychological research ever performed. Think you know what makes you happy? You won't know for sure until you have read this book'
STEVE LEVITT, author of *Freakonomics*

'In *Stumbling on Happiness*, Daniel Gilbert shares his brilliant insights into our quirks of mind, and steers us toward happiness in the most delightful, engaging ways. If you stumble on this book, you're guaranteed many doses of joy'
DANIEL GOLEMAN, author of *Emotional Intelligence*

'This is a brilliant book, a useful book, and a book that could quite possibly change the way you look at just about everything. And as a bonus, Gilbert writes like a cross between Malcolm Gladwell and David Sedaris'
SETH GODIN, author of *All Marketers Are Liars*

'Everyone will enjoy reading this book, and some of us will wish we could have written it. You will rarely have a chance to learn so much about so important a topic while having so much fun'
PROFESSOR DANIEL KAHNEMAN, Princeton University, Winner of the 2002 Nobel Prize in Economics

DANIEL GILBERT

Stumbling on Happiness

HARPER PERENNIAL
London, New York, Toronto and Sydney

Harper Perennial
An imprint of HarperCollins*Publishers*
1 London Bridge Street
London SE1 9GF

www.harperperennial.co.uk

HarperCollinsPublishers
1st Floor, Watermarque Building, Ringsend Road
Dublin 4, Ireland

This edition published by Harper Perennial 2007

20

First published in Great Britain by Harper Press in 2006

ISBN-13 978-0-00-718313-5

Set in Sabon

Printed and bound in the UK using
100% Renewable Electricity at CPI Group (UK) Ltd

For Oli, under the apple tree

One cannot divine nor forecast the conditions that will make happiness; one only stumbles upon them by chance, in a lucky hour, at the world's end somewhere, and holds fast to the days, as to fortune or fame.

Willa Cather, 'Le Lavandou', 1902

CONTENTS

Acknowledgments   xi

Foreword   xiii

PART I   PROSPECTION   1

1. Journey to Elsewhen   3

PART II   SUBJECTIVITY   27

2. The View from in Here   29
3. Outside Looking In   55

PART III   REALISM   73

4. In the Blind Spot of the Mind's Eye   75
5. The Hound of Silence   96

PART IV   PRESENTISM   109

6. The Future Is Now   111
7. Time Bombs   127

PART V   RATIONALIZATION   149

8. Paradise Glossed   151
9. Immune to Reality   172

PART VI   CORRIGIBILITY   193

10. Once Bitten   195
11. Reporting Live from Tomorrow   212

Contents

Afterword	235
Notes	239
Index	269

ACKNOWLEDGMENTS

THIS IS THE PART OF THE BOOK in which the author typically claims that nobody writes a book by himself and then names all the people who presumably wrote the book for him. It must be nice to have friends like that. Alas, all the people who wrote this book are me, so let me instead thank those who by their gifts enabled me to write a book without them.

First and foremost, I thank the students and colleagues who did so much of the research described in these pages and let me share in the credit. They include Danny Axsom, Mike Berkovits, Stephen Blumberg, Ryan Brown, David Centerbar, Erin Driver-Linn, Liz Dunn, Jane Ebert, Mike Gill, Sarit Golub, Karim Kassam, Debbie Kermer, Boaz Keysar, Jaime Kurtz, Matt Lieberman, Jay Meyers, Carey Morewedge, Kristian Myrseth, Becca Norwick, Kevin Ochsner, Liz Pinel, Jane Risen, Todd Rogers, Ben Shenoy and Thalia Wheatley. How did I get lucky enough to work with all of you?

I owe a very special debt of gratitude to my friend and longtime collaborator Tim Wilson of the University of Virginia, whose creativity and intelligence have been constant sources of inspiration, envy and research grants. The previous sentence is the only one in this book that I could possibly have written without him.

Several colleagues read chapters, made suggestions, provided information or in some other way spared the wild geese a good chasing. They include Sissela Bok, Allan Brandt, Patrick Cavanagh, Nick Epley, Nancy Etcoff, Tom Gilovich, Richard Hackman, John Helliwell, Danny Kahneman, Boaz Keysar, Jay Koehler, Steve Kosslyn, David Laibson, Andrew Oswald, Steve Pinker, Rebecca Saxe, Jonathan Schooler, Nancy Segal, Dan Simons, Robert Trivers, Dan Wegner and Tim Wilson. Thank you all.

My agent, Katinka Matson, dared me to stop yapping about this book and to start writing it, and although she isn't the only person

Acknowledgments

who ever told me to stop yapping, she's the only one I still like. My editor at Knopf, Marty Asher, has a beautiful ear and a big blue pencil, and if you don't think this book is a pleasure to read, then you should have seen it before he got ahold of it.

I wrote much of this book while on sabbatical leaves that were subsidized by the President and Fellows of Harvard College, the John Simon Guggenheim Memorial Foundation, the James McKeen Cattell Foundation, the American Philosophical Society, the National Institute of Mental Health and the University of Chicago Graduate School of Business. I thank these institutions for investing in my disappearance.

And finally, the mush. I am grateful for the coincidence of having a wife and a best friend who are both named Marilynn Oliphant. No one should have to pretend to be interested in every half-baked thought that pops into my head. No one should, but someone does. The members of the Gilbert and Oliphant clans – Larry, Gloria, Sherry, Scott, Diana, Mister Mikey, Jo, Danny, Shona, Arlo, Amanda, Big Z, Sarah B., Wren and Daylyn – share joint custody of my heart, and I thank them all for giving that heart a home. Finally, allow me to remember with gratitude and affection two souls whom even heaven does not deserve: my mentor, Ned Jones, and my mother, Doris Gilbert.

Now let's go stumbling.

18 July 2005
Cambridge, Massachusetts

FOREWORD

How sharper than a serpent's tooth it is
To have a thankless child.

Shakespeare, *King Lear*

WHAT WOULD YOU DO right now if you learned that you were going to die in ten minutes? Would you race upstairs and light that Marlboro you've been hiding in your sock drawer since the Ford administration? Would you waltz into your boss's office and present him with a detailed description of his personal defects? Would you drive out to that steakhouse near the new mall and order a T-bone, medium rare, with an extra side of the really *bad* cholesterol? Hard to say, of course, but of all the things you might do in your final ten minutes, it's a pretty safe bet that few of them are things you actually did today.

Now, some people will bemoan this fact, wag their fingers in your direction, and tell you sternly that you should live every minute of your life as though it were your last, which only goes to show that some people would spend their final ten minutes giving other people dumb advice. The things we do when we expect our lives to continue are naturally and properly different than the things we might do if we expected them to end abruptly. We go easy on the lard and tobacco, smile dutifully at yet another of our boss's witless jokes, read books like this one when we could be wearing paper hats and eating pistachio macaroons in the bathtub, and we do each of these things in the charitable service of the people we will soon become. We treat our future selves as though they were our children, spending most of the hours of most of our days constructing tomorrows that we hope will make them happy. Rather than indulging in whatever strikes our momentary fancy, we take responsibility for the welfare of our future selves, squirrelling away portions of our pay-cheques each month so *they* can enjoy their retirements on a putting green, jogging and flossing with some regularity so *they* can avoid coronaries and gum grafts, enduring dirty nappies and mind-numbing repetitions of *The Cat in the Hat* so that someday *they* will have fat-cheeked grandchildren to bounce on their laps. Even plunking down a dollar at the convenience store is an act of charity intended to ensure that the person we are about to become will enjoy the cupcake we are paying for now. In fact, just about any time we *want* something – a promotion, a marriage, an automobile, a cheeseburger – we are expecting that if we get it, then the person who has our fingerprints a second, minute, day or decade from now will enjoy the world they inherit from us, honoring our sacrifices as they reap the harvest of our shrewd investment decisions and dietary forbearance.

Yeah, yeah. Don't hold your breath. Like the fruits of our loins, our temporal progeny are often thankless. We toil and sweat to give them just what we think they will like, and they quit their jobs, grow their hair, move to or from San Francisco and wonder how we could ever have been stupid enough to think they'd like *that*. We fail to achieve the accolades and rewards that we consider crucial to their well-being, and they end up thanking God that things didn't work out according to our shortsighted, misguided plan. Even that person who takes a bite of the cupcake we purchased a few minutes earlier may make a sour face and accuse *us* of having bought the wrong snack. No one likes to be criticized, of course, but if the things we successfully strive for do not make our future selves happy, or if the things we unsuccessfully avoid do, then it seems reasonable (if somewhat ungracious) for them to cast a disparaging glance backward and wonder what the hell we were thinking. They may recognize our good intentions and begrudgingly acknowledge that we did the best we could, but they will inevitably whine to their therapists about how our best just wasn't good enough for them.

How can this happen? Shouldn't we know the tastes, preferences, needs and desires of the people we will be next year – or at least later this afternoon? Shouldn't we understand our future selves well enough to shape their lives – to find careers and lovers whom they will cherish, to buy slipcovers for the sofa that they will treasure for years to come? So why do they end up with attics and lives that are full of stuff that we considered indispensable and that they consider painful, embarrassing or useless? Why do they criticize our choice of romantic partners, second-guess our strategies for professional advancement and pay good money to remove the tattoos that we paid good money to get? Why do they experience regret and relief when they think about us, rather than pride and appreciation? We might understand all this if we had neglected them, ignored them, mistreated them in some fundamental way – but damn it, we gave them the best years of our lives! How can they be disappointed when we accomplish our coveted goals, and why are they so damned *giddy* when they end up in precisely the spot that we worked so hard to steer them clear of? Is there something wrong with them?

Or is there something wrong with us?

WHEN I WAS TEN YEARS OLD, the most magical object in my house was a book on optical illusions. Its pages introduced me to the Müller-Lyer lines whose arrow-tipped ends made them appear as though they were different lengths even though a ruler showed them to be identical, the Necker cube that appeared to have an open side one moment and then an open top the next, the drawing of a chalice that suddenly became a pair of silhouetted faces before flickering back into a chalice again (see figure 1). I would sit on the floor in my father's study and stare at that book for hours, mesmerized by the fact that these simple drawings could force my brain to believe things that it knew with utter certainty to be wrong. This is when I learned that mistakes are interesting and began planning a life that contained several of them. But an optical illusion is not interesting

Fig. 1.

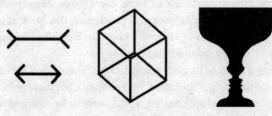

simply because it causes everyone to make a mistake; rather, it is interesting because it causes everyone to make the *same* mistake. If I saw a chalice, you saw Elvis and a friend of ours saw a paper carton of moo goo gai pan, then the object we were looking at would be a very fine inkblot but a lousy optical illusion. What is so compelling about optical illusions is that everyone sees the chalice first, the faces next, and then – *flicker flicker* – there's that chalice again. The errors that optical illusions induce in our perceptions are lawful, regular and systematic. They are not dumb mistakes but smart mistakes – mistakes that allow those who understand them to glimpse the elegant design and inner workings of the visual system.

The mistakes we make when we try to imagine our personal futures are also lawful, regular and systematic. They too have a pattern that tells us about the powers and limits of foresight in much the same way that optical illusions tell us about the powers and limits of eyesight. That's what this book is all about. Despite the third word of the title, this is not an instruction manual that will tell you anything useful about how to be happy. Those books are located in the self-help section two aisles over, and once you've bought one, done everything it says to do and found yourself miserable anyway, you can always come back here to understand why. Instead, this is a book that describes what science has to tell us about how and how well the human brain can imagine its own future, and about how and how well it can predict which of those futures it will most enjoy. This book is about a puzzle that many thinkers have pondered over the last two millennia, and it uses their ideas (and a few of my own) to explain why we seem to know so little about the hearts and minds of the people we are about to become. The story is a bit like a river that crosses borders without benefit of passport because no single science has ever produced a compelling solution to the puzzle. Weaving together facts and theories from psychology, cognitive neuroscience, philosophy and behavioural economics, this book allows an account to emerge that I personally find convincing but whose merits you will have to judge for yourself.

Writing a book is its own reward, but reading a book is a commitment of time and money that ought to pay clear dividends. If you are not educated and entertained, you deserve to be returned to your original age and net worth. That won't happen, of course, so I've written a book that I hope will interest and amuse you, provided you don't take yourself too seriously and have at least ten minutes to live. No one can say how you will feel when you get to the end of this book, and that includes the you who is about to start it. But if your future self is not satisfied when it arrives at the last page, it will at least understand why you mistakenly thought it would be.[1]

PART I

Prospection

prospection (pro·spe·kshen)
The act of looking forward in time or
considering the future.

CHAPTER I

Journey to Elsewhen

> O, that a man might know
> The end of this day's business ere it come!
>
> Shakespeare, *Julius Caesar*

PRIESTS VOW TO REMAIN CELIBATE, physicians vow to do no harm and letter carriers vow to swiftly complete their appointed rounds despite snow, sleet and split infinitives. Few people realize that psychologists also take a vow, promising that at some point in their professional lives they will publish a book, a chapter or at least an article that contains this sentence: 'The human being is the only animal that . . .' We are allowed to finish the sentence any way we like, but it has to start with those eight words. Most of us wait until relatively late in our careers to fulfil this solemn obligation because we know that successive generations of psychologists will ignore all the other words that we managed to pack into a lifetime of well-intentioned scholarship and remember us mainly for how we finished *The Sentence*. We also know that the worse we do, the better we will be remembered. For instance, those psychologists who finished The Sentence with 'can use language' were particularly well remembered when chimpanzees were taught to communicate with hand signs. And when researchers discovered that chimps in the wild use sticks to extract tasty termites from their mounds (and to bash one another over the head now and then), the world suddenly remembered the full name and mailing address of every psychologist who had ever finished The Sentence with 'uses tools'. So it is for good reason that most psychologists put off completing The Sentence for as long as they can, hoping that if they wait long enough, they just might die in time to avoid being publicly humiliated by a monkey.

I have never before written The Sentence, but I'd like to do so now, with you as my witness. *The human being is the only animal that thinks about the future.* Now, let me say up front that I've had cats, I've had dogs, I've had gerbils, mice, goldfish and crabs (no, not that kind), and I do recognize that nonhuman animals often *act* as though they have the capacity to think about the future. But as bald men with cheap hairpieces always seem to forget, acting as though you have something and actually having it are not the same thing, and anyone who looks closely can tell the difference. For example, I live in an urban neighbourhood, and every autumn the squirrels in my yard (which is approximately the size of two squirrels) act as though they know that they will be unable to eat later unless they bury some food now. My city has a relatively well-educated citizenry, but as far as anyone can tell its squirrels are not particularly distinguished. Rather, they have regular squirrel brains that run food-burying programs when the amount of sunlight that enters their regular squirrel eyes decreases by a critical amount. Shortened days trigger burying behaviour with no intervening contemplation of tomorrow, and the squirrel that stashes a nut in my yard 'knows' about the future in approximately the same way that a falling rock 'knows' about the law of gravity – which is to say, not really. Until a chimp weeps at the thought of growing old alone, or smiles as it contemplates its summer holiday, or turns down a toffee apple because it already looks too fat in shorts, I will stand by my version of The Sentence. We think about the future in a way that no other animal can, does or ever has, and this simple, ubiquitous, ordinary act is a defining feature of our humanity.[1]

The Joy of Next

If you were asked to name the human brain's greatest achievement, you might think first of the impressive artefacts it has produced – the Great Pyramid of Giza, the International Space Station or perhaps the Golden Gate Bridge. These are great achievements indeed, and our brains deserve their very own ticker-tape parade for producing them. But they are not the greatest. A sophisticated machine could design and build any one of these things because designing and building require knowledge, logic and patience, of which sophisticated machines have plenty. In fact, there's really only *one* achievement so remarkable that even the most sophisticated machine cannot pretend to have accomplished it, and that achievement is conscious experience. *Seeing* the Great Pyramid or *remembering* the Golden Gate or *imagining* the Space Station are far more remarkable acts than is building any one of them. What's more, one of these remarkable acts is even more remarkable than the others. To see is to experience the world as it is, to remember is to experience the world as it was, but to imagine – ah, to *imagine* is to experience the world as it isn't and has never been, but as it might be. The greatest achievement of the human brain is its ability to imagine objects and episodes that do not exist in the realm of the real, and it is this ability that allows us to think about the future. As one philosopher noted, the human brain is an 'anticipation machine', and 'making future' is the most important thing it does.[2]

But what exactly does 'making future' mean? There are at least two ways in which brains might be said to make future, one of which we share with many other animals, the other of which we share with none. All brains – human brains, chimpanzee brains, even ordinary food-burying squirrel brains – make predictions about the *immediate, local, personal future*. They do this by using information about current events ('I smell something') and past events ('Last time I smelled this smell, a big thing tried to eat me') to anticipate the event that is most likely to happen to them next ('A big thing is about to ———').[3] But notice two features of this so-called prediction. First, despite the comic quips inside the parentheses, predictions such as these do not require the brain making them to have anything even remotely resembling a conscious thought. Just as an abacus can put two and two together to produce four without having thoughts about arithmetic, so brains can add past to present to make future without ever thinking about any of them. In fact, it doesn't even require a brain to make predictions such as these. With just a little bit of training, the giant sea slug known as *Aplysia parvula* can learn to predict and avoid an electric shock to its gill, and as anyone with a scalpel can easily demonstrate, sea slugs are inarguably brainless. Computers are also brainless, but they use precisely the same trick the sea slug does when they turn down your credit card because you were trying to buy dinner in Paris after buying lunch in LA. In short, machines and invertebrates prove that it doesn't take a smart, self-aware, conscious brain to make simple predictions about the future.

The second thing to notice is that predictions such as these are not particularly far-reaching. They are not predictions in the same sense that we might predict the annual rate of inflation, the intellectual impact of postmodernism, the heat death of the universe, or Madonna's next hair colour. Rather, these are predictions about what will happen in precisely this spot, precisely next, to precisely me, and we call them *predictions* only because there is no better word for them in the English language. But the use of that term – with its inescapable connotations of calculated, thoughtful reflection about events that may occur anywhere, to anyone, at any time – risks obscuring the fact that brains are continuously making predictions about the immediate, local, personal future of their owners without their owners' awareness. Rather than saying that such brains are *predicting* let's say that they are *nexting*.

Yours is nexting right now. For example, at this moment you may be consciously thinking about the sentence you just read, or about the key ring in your pocket that is jammed uncomfortably against your thigh, or about whether the War of 1812 really deserves its own overture. Whatever you are thinking, your thoughts are surely about something other than the word with which this sentence will end. But even as you hear these very words echoing in your very head, and think whatever thoughts they inspire, your brain is using the word it is reading *right now* and the words it read *just before* to make a reasonable guess about the identity of the word it will read *next* which is what allows you to read so fluently.[4] Any brain that has been raised on a steady diet of film noir and cheap detective novels fully expects the word *night* to follow the phrase *It was a dark and stormy* and thus when it does encounter the word *night* it is especially well prepared to digest it. As long as your brain's guess about the next word turns out to be right, you cruise along happily, left to right, left to right, turning black squiggles into ideas, scenes, characters and concepts, blissfully unaware that your nexting brain is predicting the future of the sentence at a fantastic rate. It is only when your brain predicts badly that you suddenly feel avocado.

That is, surprised. See?

Now, consider the meaning of that brief moment of surprise. Surprise is an emotion we feel when we encounter the unexpected – for example, thirty-four acquaintances in paper hats standing in our living room yelling 'Happy birthday!' as we walk through the front door with a bag of groceries and a full bladder – and thus the occurrence of surprise reveals the nature of our expectations. The surprise you felt at the end of the last paragraph reveals that as you were reading the phrase *it is only when your brain predicts badly that you suddenly feel . . .* , your brain was simultaneously making a reasonable prediction about what would happen next. It predicted that sometime in the next few milliseconds your eyes would come across a set of black squiggles that encoded an English word that described a feeling, such as *sad* or *nauseous* or even *surprised*. Instead, it encountered a fruit, which woke you from your dogmatic slumbers and revealed the nature of your expectations to anyone who was watching. Surprise tells us that we were expecting something other than what we got, even when we didn't know we were expecting anything at all.

Because feelings of surprise are generally accompanied by reactions that can be observed and measured – such as eyebrow arching, eye widening, jaw dropping, and noises followed by a series of exclamation marks – psychologists can use surprise to tell them when a brain is nexting. For example, when monkeys see a researcher drop a ball down one of several chutes, they quickly look to the bottom of that chute and wait for the ball to reemerge. When some experimental trickery causes the ball to emerge from a different chute than the one in which it was deposited, the monkeys display surprise, presumably because their brains were nexting.[5] Human babies have similar responses to weird physics. For example, when babies are shown a video of a big red block smashing into a little yellow block, they react with indifference when the little yellow block instantly goes careening off the screen. But when the little yellow block hesitates for just a moment or two before careening away, babies stare like bystanders at a train wreck – as though the delayed careening had violated some prediction made by their nexting brains.[6] Studies such as these tell us that monkey brains 'know' about gravity (objects fall down, not sideways) and that baby human brains 'know' about kinetics (moving objects transfer energy to stationary objects at precisely the moment they contact them and not a few seconds later). But more important, they tell us that monkey brains and baby human brains add what they already know (the past) to what they currently see (the present) to predict what will happen next (the future). When the actual next thing is different from the predicted next thing, monkeys and babies experience surprise.

Our brains were made for nexting, and that's just what they'll do. When we take a stroll on the beach, our brains predict how stable the sand will be when our foot hits it, and then adjust the tension in our knee accordingly. When we leap to catch a Frisbee, our brains predict where the disc will be when we cross its flight path, and then bring our hands to precisely that point. When we see a sand crab scurry behind a bit of driftwood on its way to the water, our brains predict when and where the critter will reappear, and then direct our eyes to the precise point of its reemergence. These predictions are remarkable in both the speed and accuracy with which they are made, and it is difficult to imagine what our lives would be like if our brains quit making them, leaving us completely 'in the moment' and unable to take our next step. But while these automatic, continuous, nonconscious predictions of the immediate, local, personal future are both amazing and ubiquitous, they are not the sorts of predictions that got our species out of the trees and into dress slacks. In fact, these are the kinds of predictions that frogs make without ever leaving their lily pads, and hence not the sort that The Sentence was meant to describe. No, the variety of future that we human beings manufacture – and that *only* we manufacture – is of another sort entirely.

The Ape That Looked Forward

Adults love to ask children idiotic questions so that we can chuckle when they give us idiotic answers. One particularly idiotic question we like to ask children is this: 'What do you want to be when you grow up?' Small children look appropriately puzzled, worried perhaps that our question implies they are at some risk of growing down. If they answer at all, they generally come up with things like 'the candy guy' or 'a tree climber'. We chuckle because the odds that the child will ever become the candy guy or a tree climber are vanishingly small, and they are vanishingly small because these are not the sorts of things that most children will want to be once they are old enough to ask idiotic questions themselves. But notice that while these are the wrong answers to our question, they are the right answers to another question, namely, 'What do you want to be *now*?' Small children cannot say what they want to be *later* because they don't really understand what *later* means.[7] So, like shrewd politicians, they ignore the question they are asked and answer the question they can. Adults do much better, of course. When a thirtyish Manhattanite is asked where she thinks she might retire, she mentions Miami, Phoenix or some other hotbed of social rest. She may love her gritty urban existence right now, but she can imagine that in a few decades she will value bingo and prompt medical attention more than art museums and squeegee men. Unlike the child who can only think about how things are, the adult is able to think about how things will be. At some point between our high chairs and our rocking chairs, we learn about later.[8]

Later! What an astonishing idea. What a powerful concept. What a fabulous discovery. How did human beings ever learn to preview in their imaginations chains of events that had not yet come to pass? What prehistoric genius first realized that he could escape today by closing his eyes and silently transporting himself into tomorrow? Unfortunately, even big ideas leave no fossils for carbon dating, and thus the natural history of *later* is lost to us forever. But paleontologists and neuroanatomists assure us that this pivotal moment in the drama of human evolution happened sometime within the last 3 million years, and that it happened quite suddenly. The first brains appeared on earth about 500 million years ago, spent a leisurely 430 million years or so evolving into the brains of the earliest primates, and another 70 million years or so evolving into the brains of the first protohumans. Then something happened – no one knows quite what, but speculation runs from the weather turning chilly to the invention of cooking – and the soon-to-be-human brain experienced an unprecedented growth spurt that more than doubled its mass in a little over two million years, transforming it from the one-and-a-quarter-pound brain of *Homo habilis* to the nearly three-pound brain of *Homo sapiens*.[9]

Now, if you were put on a hot-fudge diet and managed to double your mass in a very short time, we would not expect all of your various body parts to share equally in the gain. Your belly and buttocks would probably be the major recipients of newly acquired flab, while your tongue and toes would remain relatively svelte and unaffected. Similarly, the dramatic increase in the size of the human brain did not democratically double the mass of every part so that modern people ended up with new brains that were structurally identical to the old ones, only bigger. Rather, a disproportionate share of the growth centred on a particular part of the brain known as the frontal lobe, which, as its name implies, sits at the front of the head, squarely above the eyes (see figure 2). The low, sloping brows of our earliest ancestors were pushed forward to become the sharp, vertical brows that keep our hats on, and the change in the structure of our heads occurred primarily to accommodate this sudden change in the size of our brains. What did this new bit of cerebral apparatus do to justify an architectural overhaul of the human skull? What is it about this particular part that made nature so anxious for each of us to have a big one? Just what good is a frontal lobe?

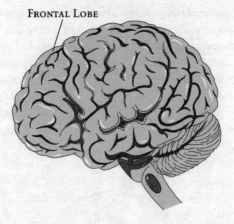

FRONTAL LOBE

Fig. 2. The frontal lobe is the recent addition to the human brain that allows us to imagine the future.

Until fairly recently, scientists thought it was not much good at all, because people whose frontal lobes were damaged seemed to do pretty well without them. Phineas Gage was a foreman for the Rutland Railroad who, on a lovely autumn day in 1848, ignited a small explosion in the vicinity of his feet, launching a three-and-a-half-foot-long iron rod into the air, which Phineas cleverly caught with his face. The rod entered just beneath his left cheek and exited through the top of his skull, boring a tunnel through his cranium and taking a good chunk of frontal lobe with it (see figure 3). Phineas was knocked to the ground, where he lay for a few minutes. Then, to everyone's astonishment, he stood up and asked if a coworker might escort him to the doctor, insisting all the while that he didn't need a ride and could walk by himself, thank you. The doctor cleaned some dirt from his wound, a coworker cleaned some brain from the rod, and in a relatively short while, Phineas and his rod were back about their business.[10] His personality took a decided turn for the worse – and that fact is the source of his fame to this day – but the more striking thing about Phineas was just how *normal* he otherwise was. Had the rod made hamburger of another brain part – the visual cortex, Broca's area, the brain stem – then Phineas might have died, gone blind, lost the ability to speak or spent the rest of his life doing a convincing impression of a cabbage. Instead, for the next twelve years, he lived, saw, spoke, worked and travelled so uncabbagely that neurologists could only conclude that the frontal lobe did little for a fellow that he couldn't get along nicely without.[11] As one neurologist wrote in 1884, 'Ever since the occurrence of the famous American crowbar case it has been known that destruction of these lobes does not necessarily give rise to any symptoms.'[12]

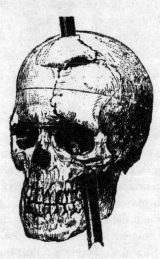

Fig. 3. An early medical sketch showing where the tamping iron entered and exited Phineas Gage's skull.

But the neurologist was wrong. In the nineteenth century, knowledge of brain function was based largely on the observation of people who, like Phineas Gage, were the unfortunate subjects of one of nature's occasional and inexact neurological experiments. In the twentieth century, surgeons picked up where nature left off and began to do more precise experiments whose results painted a very different picture of frontal lobe function. In the 1930s, a Portuguese physician named António Egas Moniz was looking for a way to quiet his highly agitated psychotic patients when he heard about a new surgical procedure called frontal lobotomy, which involved the chemical or mechanical destruction of parts of the frontal lobe. This procedure had been performed on monkeys, who were normally quite angry when their food was withheld, but who reacted to such indignities with unruffled patience after experiencing the operation. Egas Moniz tried the procedure on his human patients and found that it had a similar calming effect. (It also had the calming effect of winning Egas Moniz the Nobel Prize for Medicine in 1949.) Over the next few decades, surgical techniques were improved (the procedure could be performed under local anesthesia with an ice pick) and unwanted side effects (such as lowered intelligence and bed-wetting) were diminished. The destruction of some part of the frontal lobe became a standard treatment for cases of anxiety and depression that resisted other forms of therapy.[13] Contrary to the conventional medical wisdom of the previous century, the frontal lobe did make a difference. The difference was that some people seemed better off without it.

But while some surgeons were touting the benefits of frontal lobe damage, others were noticing the costs. Although patients with frontal lobe damage often performed well on standard intelligence tests, memory tests and the like, they showed severe impairments on any test – even the very simplest test – that involved planning. For instance, when given a maze or a puzzle whose solution required that they consider an entire series of moves before making their first move, these otherwise intelligent people were stumped.[14] Their planning deficits were not limited to the laboratory. These patients might function reasonably well in ordinary situations, drinking tea without spilling and making small talk about the curtains, but they found it practically impossible to say what they would do later that afternoon. In summarizing scientific knowledge on this topic, a prominent scientist concluded: 'No prefrontal symptom has been reported more consistently than the inability to plan. . . . The symptom appears unique to dysfunction of the prefrontal cortex . . . [and] is not associated with clinical damage to any other neural structure.'[15]

Now, this pair of observations – that damage to certain parts of the frontal lobe can make people feel calm but that it can also leave them unable to plan – seem to converge on a single conclusion. What is the conceptual tie that binds *anxiety* and *planning*? Both, of course, are intimately connected to thinking about the future. We feel anxiety when we anticipate that something bad will happen, and we plan by imagining how our actions will unfold over time. Planning requires that we peer into our futures, and anxiety is one of the reactions we may have when we do.[16] The fact that damage to the frontal lobe impairs planning and anxiety so uniquely and precisely suggests that the frontal lobe is the critical piece of cerebral machinery that allows normal, modern human adults to project themselves into the future. Without it we are trapped in the moment, unable to imagine tomorrow and hence unworried about what it may bring. As scientists now recognize, the frontal lobe 'empowers healthy human adults with the capacity to consider the self's extended existence throughout time'.[17] As such, people whose frontal lobe is damaged are described by those who study them as being 'bound to present stimuli',[18] or 'locked into immediate space and time',[19] or as displaying a 'tendency toward temporal concreteness'.[20] In other words, like candy guys and tree climbers, they live in a world without *later*.

The sad case of the patient known as N.N. provides a window into this world. N.N. suffered a closed head injury in an automobile accident in 1981, when he was thirty years old. Tests revealed that he had sustained extensive damage to his frontal lobe. A psychologist interviewed N.N. a few years after the accident and recorded this conversation:

PSYCHOLOGIST: What will you be doing tomorrow?
N.N.: I don't know.
PSYCHOLOGIST: Do you remember the question?
N.N.: About what I'll be doing tomorrow?
PSYCHOLOGIST: Yes, would you describe your state of mind when you try to think about it?
N.N.: Blank, I guess . . . It's like being asleep . . . like being in a room with nothing there and having a guy tell you to go find a chair, and there's nothing there . . . like swimming in the

middle of a lake. There's nothing to hold you up or do anything with.[21]

N.N.'s inability to think about his own future is characteristic of
patients with frontal lobe damage. For N.N., tomorrow will always
be an empty room, and when he attempts to envision *later*, he will
always feel as the rest of us do when we try to imagine nonexistence
or infinity. Yet, if you struck up a conversation with N.N. on the
subway, or chatted with him while standing in a queue at the post
office, you might not know that he was missing something so fundamentally human. After all, he understands *time* and *the future* as
abstractions. He knows what hours and minutes are, how many of
the latter there are in the former, and what *before* and *after* mean. As
the psychologist who interviewed N.N. reported: "He knows many
things about the world, he is aware of this knowledge, and he can
express it flexibly. In this sense he is not greatly different from a normal adult. But he seems to have no capacity of experiencing extended
subjective time. . . . He seems to be living in a 'permanent present.'"[22]

A permanent present – what a haunting phrase. How bizarre and
surreal it must be to serve a life sentence in the prison of the moment,
trapped forever in the perpetual now, a world without end, a time
without later. Such an existence is so difficult for most of us to imagine, so alien to our normal experience, that we are tempted to dismiss it as a fluke – an unfortunate, rare and freakish aberration
brought on by traumatic head injury. But in fact, this strange existence is the rule and *we* are the exception. For the first few hundred
million years after their initial appearance on our planet, all brains
were stuck in the permanent present, and most brains still are today.
But not yours and not mine, because two or three million years ago
our ancestors began a great escape from the here and now, and their
getaway vehicle was a highly specialized mass of grey tissue, fragile,
wrinkled and appended. This frontal lobe – the last part of the
human brain to evolve, the slowest to mature and the first to deteriorate in old age – is a time machine that allows each of us to vacate
the present and experience the future before it happens. No other
animal has a frontal lobe quite like ours, which is why we are the

only animal that thinks about the future as we do. But if the story of the frontal lobe tells us *how* people conjure their imaginary tomorrows, it doesn't tell us *why*.

Twisting Fate

In the late 1960s, a Harvard psychology professor took LSD, resigned his appointment (with some encouragement from the administration), went to India, met a guru and returned to write a popular book called *Be Here Now*, whose central message was succinctly captured by the injunction of its title.[23] The key to happiness, fulfilment and enlightenment, the ex-professor argued, was to stop thinking so much about the future.

Now, why would anyone go all the way to India and spend his time, money and brain cells just to learn how not to think about the future? Because, as anyone who has ever tried to learn meditation knows, not thinking about the future is much more challenging than being a psychology professor. Not to think about the future requires that we convince our frontal lobe not to do what it was designed to do, and like a heart that is told not to beat, it naturally resists this suggestion. Unlike N.N., most of us do not *struggle* to think about the future because mental simulations of the future arrive in our consciousness regularly and unbidden, occupying every corner of our mental lives. When people are asked to report how much they think about the past, present and future, they claim to think about the future the most.[24] When researchers actually *count* the items that float along in the average person's stream of consciousness, they find that about 12 per cent of our daily thoughts are about the future.[25] In other words, every eight hours of thinking includes an hour of thinking about things that have yet to happen. If you spent one out of every eight hours living in my state you would be required to pay taxes, which is to say that in some very real sense, each of us is a part-time resident of tomorrow.

Why can't we just be here now? How come we can't do something our goldfish find so simple? Why do our brains stubbornly insist on projecting us into the future when there is so much to think about right here today?

Prospection and Emotion

The most obvious answer to that question is that thinking about the future can be pleasurable. We daydream about hitting a home-run at the company picnic, posing with the lottery commissioner and the door-sized cheque, or making snappy patter with the attractive teller at the bank – not because we expect or even want these things to happen, but because merely imagining these possibilities is itself a source of joy. Studies confirm what you probably suspect: when people daydream about the future, they tend to imagine themselves achieving and succeeding rather than fumbling or failing.[26]

Indeed, thinking about the future can be so pleasurable that sometimes we'd rather think about it than get there. In one study, volunteers were told that they had won a free dinner at a fabulous French restaurant and were then asked when they would like to eat it. Now? Tonight? Tomorrow? Although the delights of the meal were obvious and tempting, most of the volunteers chose to put their restaurant visit off a bit, generally until the following week.[27] Why the self-imposed delay? Because by waiting a week, these people not only got to spend several hours slurping oysters and sipping Château Cheval Blanc '47, but they also got to look forward to all that slurping and sipping for a full seven days beforehand. Forestalling pleasure is an inventive technique for getting double the juice from half the fruit. Indeed, some events are more pleasurable to imagine than to experience (most of us can recall an instance in which we made love with a desirable partner or ate a wickedly rich dessert, only to find that the act was better contemplated than consummated), and in these cases people may decide to delay the event forever. For instance, volunteers in one study were asked to imagine themselves requesting a date with a person on whom they had a major crush, and those who had had the most elaborate and delicious fantasies about approaching their heartthrob were *least* likely to do so over the next few months.[28]

We like to frolic in the best of all imaginary tomorrows – and why shouldn't we? After all, we fill our photo albums with pictures of birthday parties and tropical holidays rather than car wrecks and emergency-room visits because we want to be happy when we stroll down Memory Lane, so why shouldn't we take the same attitude toward our strolls up Imagination Avenue? Although imagining happy futures may make us feel happy, it can also have some troubling consequences. Researchers have discovered that when people find it easy to imagine an event, they overestimate the likelihood that it will actually occur.[29] Because most of us get so much more practice imagining good than bad events, we tend to overestimate the likelihood that good events will actually happen to us, which leads us to be unrealistically optimistic about our futures.

For instance, American college students expect to live longer, stay married longer and travel to Europe more often than average.[30] They believe they are more likely to have a gifted child, to own their own home and to appear in the newspaper, and less likely to have a heart attack, venereal disease, a drinking problem, an auto accident, a broken bone or gum disease. Americans of all ages expect their futures to be an improvement on their presents,[31] and although citizens of other nations are not quite as optimistic as Americans, they also tend to imagine that their futures will be brighter than those of their peers.[32] These overly optimistic expectations about our personal futures are not easily undone: experiencing an earthquake causes people to become temporarily realistic about their risk of dying in a future disaster, but within a couple of weeks even earthquake survivors return to their normal level of unfounded optimism.[33] Indeed, events that challenge our optimistic beliefs can sometimes make us *more* rather than *less* optimistic. One study found that cancer patients were *more* optimistic about their futures than were their healthy counterparts.[34]

Of course, the futures that our brains insist on simulating are not all wine, kisses and tasty bivalves. They are often mundane, irksome, stupid, unpleasant or downright frightening, and people who seek treatment for their inability to stop thinking about the future are usually *worrying* about it rather than revelling in it. Just as a loose tooth seems to beg for wiggling, we all seem perversely compelled to imagine disasters and tragedies from time to time. On the way to the airport we imagine a future scenario in which the plane takes off without us and we miss the important meeting with the client. On the way to the dinner party we imagine a future scenario in which everyone hands the hostess a bottle of wine while we greet her empty-handed and embarrassed. On the way to the medical centre we imagine a future scenario in which our doctor inspects our chest X-ray, frowns and says something ominous such as 'Let's talk about your options.' These dire images make us feel dreadful – quite literally – so why do we go to such great lengths to construct them?

Two reasons. First, anticipating unpleasant events can minimize their impact. For instance, volunteers in one study received a series of twenty electric shocks and were warned three seconds before the onset of each one.[35] Some volunteers (the high-shock group) received twenty high-intensity shocks to their right ankles. Other volunteers (the low-shock group) received three high-intensity shocks and seventeen low-intensity shocks. Although the low-shock group received fewer volts than the high-shock group did, their hearts beat faster, they sweated more profusely and they rated themselves as more afraid. Why? Because volunteers in the low-shock group received shocks of different intensities at different times, which made it impossible for them to anticipate their futures. Apparently, three big jolts that one cannot foresee are more painful than twenty big jolts that one can.[36]

The second reason why we take such pains to imagine unpleasant events is that fear, worry and anxiety have useful roles to play in our lives. We motivate employees, children, spouses and pets to do the right thing by dramatizing the unpleasant consequences of their misbehaviours, and so too do we motivate ourselves by imagining the unpleasant tomorrows that await us should we decide to go light on the sunscreen and heavy on the éclairs. Forecasts can be 'fearcasts'[37] whose purpose is not to predict the future so much as to preclude it, and studies have shown that this strategy is often an effective way to motivate people to engage in prudent, prophylactic behavior.[38] In short, we sometimes imagine dark futures just to scare our own pants off.

Prospection and Control

Prospection can provide pleasure and prevent pain, and this is one of the reasons why our brains stubbornly insist on churning out thoughts of the future. But it is not the most important reason. Americans gladly pay millions – perhaps even billions – of dollars every year to psychics, investment advisors, spiritual leaders, weather forecasters and other assorted hucksters who claim they can predict the future. Those of us who subsidize these fortune-telling industries do not want to know what is likely to happen just for the joy of anticipating it. We want to know what is likely to happen so that we can *do something* about it. If interest rates are going to skyrocket next month, then we want to shift our money out of bonds right now. If it is going to rain this afternoon, then we want to grab an umbrella this morning. Knowledge is power, and the most important reason why our brains insist on simulating the future even when we'd rather be here now, enjoying a goldfish moment, is that our brains want to *control* the experiences we are about to have.

But why should we want to have control over our future experiences? On the face of it, this seems about as nonsensical as asking why we should want to have control over our television sets and our automobiles. But indulge me. We have a large frontal lobe so that we can look into the future, we look into the future so that we can make predictions about it, we make predictions about it so that we can control it – but why do we want to control it at all? Why not just let the future unfold as it will and experience it as it does? Why not be *here* now and *there* then? There are two answers to this question, one of which is surprisingly right and the other of which is surprisingly wrong.

The surprisingly right answer is that people find it gratifying to *exercise* control – not just for the futures it buys them, but for the exercise itself. Being effective – changing things, influencing things, making things happen – is one of the fundamental needs with which human brains seem to be naturally endowed, and much of our behavior from infancy onward is simply an expression of this penchant for control.[39] Before our butts hit the very first nappy, we already have a throbbing desire to suck, sleep, poo and make things happen. It takes us a while to get around to fulfilling the last of these desires only because it takes us a while to figure out that we have fingers, but when we do, look out world. Toddlers squeal with delight when they knock over a stack of blocks, push a ball or squash a cupcake on their foreheads. Why? Because *they did it, that's why. Look, Mum, my hand made that happen. The room is different because I was in it. I thought about falling blocks, and poof, they fell.*

The fact is that human beings come into the world with a passion for control, they go out of the world the same way, and research suggests that if they lose their ability to control things at any point between their entrance and their exit, they become unhappy, helpless, hopeless and depressed.[40] And occasionally dead. In one study, researchers gave elderly residents of a local nursing home a houseplant. They told half the residents that they were in control of the plant's care and feeding (high-control group), and they told the remaining residents that a staff person would take responsibility for the plant's well-being (low-control group).[41] Six months later, 30 per cent of the residents in the low-control group had died, compared with only 15 per cent of the residents in the high-control group. A follow-up study confirmed the importance of perceived control for the welfare of nursing-home residents but had an unexpected and unfortunate end.[42] Researchers arranged for student volunteers to pay regular visits to nursing-home residents. Residents in the high-control group were allowed to control the timing and duration of the student's visit ('Please come visit me next Thursday for an hour'), and residents in low-control group were not ('I'll come visit you next Thursday for an hour'). After two months, residents in the high-control group were happier, healthier, more active and taking fewer medications than those in the low-control group. At this point the researchers concluded their study and discontinued the student visits. Several months later they were chagrined to learn that a disproportionate number of residents who had been in the high-control group had died. Only in retrospect did the cause of this tragedy seem clear. The residents who had been given control, and who had benefited measurably from that control while they had it, were inadvertently robbed of control when the study ended.

Apparently, gaining control can have a positive impact on one's health and well-being, but losing control can be worse than never having had any at all.

Our desire to control is so powerful, and the feeling of being in control so rewarding, that people often act as though they can control the uncontrollable. For instance, people bet more money on games of chance when their opponents seem incompetent than competent – as though they believed they could control the random drawing of cards from a deck and thus take advantage of a weak opponent.[43] People feel more certain that they will win a lottery if they can control the number on their ticket,[44] and they feel more confident that they will win a dice toss if they can throw the dice themselves.[45] People will wager more money on dice that have not yet been tossed than on dice that have already been tossed but whose outcome is not yet known,[46] and they will bet more if they, rather than someone else, are allowed to decide which number will count as a win.[47] In each of these instances, people behave in a way that would be utterly absurd if they believed that they had no control over an uncontrollable event. But if somewhere deep down inside they believed that they *could* exert control – even one smidgen of an iota of control – then their behavior would be perfectly reasonable. And deep down inside, that's precisely what most of us seem to believe. Why isn't it fun to watch a videotape of last night's football game even when we don't know who won? Because the fact that the game has already been played precludes the possibility that our cheering will somehow penetrate the television, travel through the cable system, find its way to the stadium, and influence the trajectory of the ball as it hurtles toward the goalposts! Perhaps the strangest thing about this illusion of control is not that it happens but that it seems to confer many of the psychological benefits of genuine control. In fact, the one group of people who seem generally immune to this illusion are the clinically depressed,[48] who tend to estimate accurately the degree to which they can control events in most situations.[49] These and other findings have led some researchers to conclude that the feeling of control – whether real or illusory – is one of the wellsprings of mental health.[50] So if the question is 'Why should we want to control our futures?' then the surprisingly right answer is that it feels good to do so – period. Impact is rewarding. Mattering makes us happy. The act of steering one's boat down the river of time is a source of pleasure, regardless of one's port of call.

Now, at this point you probably believe two things. First, you probably believe that if you never heard the phrase "the river of time" again, it would be too soon. Amen. Second, you probably believe that even if the act of steering a metaphorical boat down a clichéd river is a source of pleasure and well-being, *where* the boat goes matters much, much more. Playing captain is a joy all its own, but the real reason why we want to steer our ships is so that we can get them to Hanalei instead of Jersey City. The nature of a place determines how we feel upon arrival, and our uniquely human ability to think about the extended future allows us to choose the best destinations and avoid the worst. We are the apes that learned to look forward because doing so enables us to shop among the many fates that might befall us and select the best one. Other animals must *experience* an event in order to learn about its pleasures and pains, but our powers of foresight allow us to imagine that which has not yet happened and hence spare ourselves the hard lessons of experience. We needn't reach out and touch an ember to know that it will hurt to do so, and we needn't experience abandonment, scorn, eviction, demotion, disease or divorce to know that all of these are undesirable ends that we should do our best to avoid. We want – and we *should* want – to control the direction of our boat because some futures are better than others, and even from this distance we should be able to tell which are which.

This idea is so obvious that it barely seems worth mentioning, but I'm going to mention it anyway. Indeed, I am going to spend the rest of this book mentioning it because it will probably take more than a few mentions to convince you that what looks like an obvious idea is, in fact, the surprisingly wrong answer to our question. We insist on steering our boats because we think we have a pretty good idea of where we should go, but the truth is that much of our steering is in vain – not because the boat won't respond, and not because we can't find our destination, but because the future is fundamentally different than it appears through the prospectiscope. Just as we experience illusions of eyesight ('Isn't it strange how one queue looks longer than the other even though it isn't?') and illusions of hindsight ('Isn't it strange how I can't remember taking out the garbage even though I did?'), so too do we experience illusions of foresight – and all three types of illusion are explained by the same basic principles of human psychology.

Onward

To be perfectly honest, I won't just be *mentioning* the surprisingly wrong answer; I'll be pounding and pummelling it until it gives up and goes home. The surprisingly wrong answer is apparently so sensible and so widely believed that only a protracted thrashing has any hope of expunging it from our conventional wisdom. So before the grudge match begins, let me share with you my plan of attack.

- In Part II, 'Subjectivity', I will tell you about the science of happiness. We all steer ourselves toward the futures that we think will make us happy, but what does that word really mean? And how can we ever hope to achieve solid, scientific answers to questions about something as gossamer as a feeling?
- We use our eyes to look into space and our imaginations to look into time. Just as our eyes sometimes lead us to see things as they are not, our imaginations sometimes lead us to foresee things as they will not be. Imagination suffers from three shortcomings that give rise to the illusions of foresight with which this book is chiefly concerned. In Part III, 'Realism', I will tell you about the first shortcoming: imagination works so quickly, quietly, and effectively that we are insufficiently sceptical of its products.
- In Part IV, 'Presentism', I will tell you about the second shortcoming: Imagination's products are . . . well, not particularly imaginative, which is why the imagined future often looks so much like the actual present.
- In Part V, 'Rationalization', I will tell you about the third shortcoming: imagination has a hard time telling us how we will *think* about the future when we get there. If we have trou-

ble foreseeing future events, then we have even more trouble foreseeing how we will see them when they happen.

- Finally, in Part VI, 'Corrigibility', I will tell you why illusions of foresight are not easily remedied by personal experience or by the wisdom we inherit from our grandmothers. I will conclude by telling you about a simple remedy for these illusions that you will almost certainly not accept.

By the time you finish these chapters, I hope you will understand why most of us spend so much of our lives turning rudders and hoisting sails, only to find that Shangri-la isn't what and where we thought it would be.

PART II

Subjectivity

subjectivity (sub·dzèk·ti·vĭtee)
The fact that experience is unobservable to
everyone but the person having it.

CHAPTER 2

The View from in Here

But, O, how bitter a thing it is to look into happiness
through another man's eyes!

Shakespeare, *As You Like It*

LORI AND REBA SCHAPPEL may be twins, but they are very different people. Reba is a somewhat shy teetotaler who has recorded an award-winning album of country music. Lori, who is outgoing, wisecracking and rather fond of strawberry daiquiris, works in a hospital and wants someday to marry and have children. They occasionally argue, as sisters do, but most of the time they get on well, complimenting each other, teasing each other and finishing each other's sentences. In fact, there are just two unusual things about Lori and Reba. The first is that they share a blood supply, part of a skull, and some brain tissue, having been joined at the forehead since birth. One side of Lori's forehead is attached to one side of Reba's, and they have spent every moment of their lives locked together, face-to-face. The second unusual thing about Lori and Reba is that they are happy – not merely resigned or contented, but joyful, playful and optimistic.[1] Their unusual life presents many challenges, of course, but as they often note, whose doesn't? When asked about the possibility of undergoing surgical separation, Reba speaks for both of them: 'Our point of view is no, straight out no. Why would you want to do that? For all the money in China, why? You'd be ruining two lives in the process.'[2]

So here's the question: if this were your life rather than theirs, how would *you* feel? If you said, 'Joyful, playful and optimistic,' then you are not playing the game and I am going to give you another chance. Try to be honest instead of correct. The honest answer is 'Despondent, desperate and depressed'. Indeed, it seems clear that no right-minded person could *really* be happy under such circumstances, which is why the conventional medical wisdom has it that conjoined twins should be separated at birth, even at the risk of killing one or both. As a prominent medical historian wrote: 'Many singletons, especially surgeons, find it inconceivable that life is worth living as a conjoined twin, inconceivable that one would not be willing to risk all – mobility, reproductive ability, the life of one or both twins – to try for separation.'³ In other words, not only does everyone know that conjoined twins will be dramatically less happy than normal people, but everyone also knows that conjoined lives are so utterly worthless that dangerous separation surgeries are an ethical imperative. And yet, standing against the backdrop of our certainty about these matters are the twins themselves. When we ask Lori and Reba how they feel about their situation, they tell us that they wouldn't have it any other way. In an exhaustive search of the medical literature, the same medical historian found the 'desire to remain together to be so widespread among communicating conjoined twins as to be practically universal'.⁴ Something is terribly wrong here. But what?

There seem to be just two possibilities. Someone – either Lori and Reba, or everyone else in the world – is making a dreadful mistake when they talk about happiness. Because we are the everyone else in question, it is only natural that we should be attracted to the former conclusion, dismissing the twins' claim to happiness with offhand rejoinders such as 'Oh, they're just saying that' or 'They may think they're happy, but they're not' or the ever popular 'They don't know what happiness really is' (usually spoken as if we do). Fair enough. But like the claims they dismiss, these rejoinders are also claims – scientific claims and philosophical claims – that presume answers to questions that have vexed scientists and philosophers for millennia. What are we all *talking* about when we make such claims about happiness?

Dancing About Architecture

There are thousands of books on happiness, and most of them start by asking what happiness *really* is. As readers quickly learn, this is approximately equivalent to beginning a pilgrimage by marching directly into the first available tar pit, because happiness *really* is nothing more or less than a word that we word makers can use to indicate anything we please. The problem is that people seem pleased to use this one word to indicate a host of different things, which has created a tremendous terminological mess on which several fine scholarly careers have been based. If one slops around in this mess long enough, one comes to see that most disagreements about what happiness *really is* are semantic disagreements about whether the word ought to be used to indicate *this* or *that*, rather than scientific or philosophical disagreements about the nature of *this* and *that*. What are the *this* and the *that* that happiness most often refers to? The word *happiness* is used to indicate at least three related things, which we might roughly call *emotional happiness*, *moral happiness*, and *judgmental happiness*.

Feeling Happy

Emotional happiness is the most basic of the trio – so basic, in fact, that we become tongue-tied when we try to define it, as though some bratty child had just challenged us to say what the word *the* means and in the process made a truly compelling case for corporal punishment. Emotional happiness is a phrase for a *feeling*, an *experience*, a *subjective state*, and thus it has no objective referent in the physical world. If we ambled down to the corner pub and met an alien from another planet who asked us to define that feeling, we would either point to the objects in the world that tend to bring it about, or we would mention other feelings that it is like. In fact, this is the only thing we *can* do when we are asked to define a subjective experience.

Consider, for instance, how we might define a very simple subjective experience, such as yellow. You may think yellow is a colour, but it isn't. It's a psychological state. It is what human beings with working visual apparatus *experience* when their eyes are struck by light with a wavelength of 580 nanometers. If our alien friend at the pub asked us to define what we were experiencing when we claimed to be *seeing yellow*, we would probably start by pointing to a mustard jar, a lemon, a rubber ducky, and saying, 'See all those things? The thing that is common to the visual experiences you have when you look at them is called *yellow*.' Or we might try to define the experience called *yellow* in terms of other experiences. 'Yellow? Well, it is sort of like the experience of orange, with a little less of the experience of red.' If the alien confided that it could not figure out what the duck, the lemon and the mustard jar had in common, and that it had never had the experience of orange or red, then it would be time to order another pint and change the topic to the universal sport of ice hockey, because there is just no other way to define yellow. Philosophers like to say that subjective states are 'irreducible', which is to say that nothing we point to, nothing we can compare them with, and nothing we can say about their neurological underpinnings can fully substitute for the experiences themselves.[5] The musician Frank Zappa is reputed to have said that writing about music is like dancing about architecture, and so it is with talking about yellow. If our new drinking buddy lacks the machinery for colour vision, then our experience of yellow is one that it will never share – or never know it shares – no matter how well we point and talk.[6]

Emotional happiness is like that. It is the feeling common to the feelings we have when we see our new granddaughter smile for the first time, receive word of a promotion, help a wayward tourist find the art museum, taste Belgian chocolate toward the back of our tongue, inhale the scent of our lover's shampoo, hear that song we used to like so much in school but haven't heard in years, touch our cheek to kitten fur, cure cancer or get a really good snootful of cocaine. These feelings are different, of course, but they also have something in common. A piece of real estate is not the same as a share of stock, which is not the same as an ounce of gold, but all are forms of *wealth* that occupy different points on a scale of *value*. Similarly, the cocaine experience is not the kitten-fur experience, which is not the promotion experience, but all are forms of *feeling* that occupy different points on a scale of *happiness*. In each of these instances, an encounter with something in the world generates a roughly similar pattern of neural activity,[7] and thus it makes sense that there is something common to our *experiences* of each – some conceptual coherence that has led human beings to group this hodgepodge of occurrences together in the same linguistic category for as long as anyone can remember. Indeed, when researchers analyse how all the words in a language are related to the others, they inevitably find that the positivity of the words – that is, the extent to which they refer to the experience of happiness or unhappiness – is the single most important determinant of their relationships.[8] Despite Tolstoy's fine efforts, most speakers consider *war* to be more closely related to *vomit* than it is to *peace*.

Happiness, then, is the you-know-what-I-mean feeling. If you are a human being who lives in this century and shares some of my cultural conditioning, then my pointing and comparing will have been effective and you will know *exactly* which feeling I mean. If you are an alien who is still struggling with yellow, then happiness is going to be a real challenge. But take heart: I would be similarly challenged if you told me that on your planet there is a feeling common to the acts of dividing numbers by three, banging one's head lightly on a doorknob, and releasing rhythmic bursts of nitrogen from any orifice at any time except on Tuesday. I would have no idea what that feeling is, and I could only learn the name and hope to use it politely in conversation. Because emotional happiness is an experience, it can only be approximately defined by its antecedents and by its relation to other experiences.[9] The poet Alexander Pope devoted about a quarter of his *Essay on Man* to the topic of happiness, and concluded with this question: 'Who thus define it, say they more or less / Than this, that happiness is happiness?'[10]

Emotional happiness may resist our efforts to tame it by description, but when we feel it, we have no doubt about its reality and its importance. Everyone who has observed human behavior for more than thirty continuous seconds seems to have noticed that people are strongly, perhaps even primarily, perhaps even single-mindedly, motivated to feel happy. If there has ever been a group of human beings who prefer despair to delight, frustration to satisfaction and pain to pleasure, they must be very good at hiding because no one has ever seen them. People want to be happy, and all the other things they want are typically meant to be means to that end. Even when people forgo happiness in the moment – by dieting when they could be eating, or working late when they could be sleeping – they are usually doing so in order to increase its future yield. The dictionary tells us that to prefer is 'to choose or want one thing rather than another *because it would be more pleasant*', which is to say that the pursuit of happiness is built into the very definition of desire. In this sense, a preference for pain and suffering is not so much a diagnosable psychiatric condition as it is an oxymoron.

Psychologists have traditionally made striving toward happiness the centrepiece of their theories of human behavior because they have found that if they don't, their theories don't work so well. As Sigmund Freud wrote:

> The question of the purpose of human life has been raised countless times; it has never yet received a satisfactory answer and perhaps does not admit of one. . . . We will therefore turn to the less ambitious question of what men show by their behavior to be the purpose and intention of their lives. What do they demand of life and wish to achieve in it? The answer to this can hardly be in doubt. They strive after happiness; they want to become happy and to remain so. This endeavour has two sides, a positive and a negative aim. It aims, on the one hand, at an absence of pain and displeasure, and, on the other, at the experiencing of strong feelings of pleasure.[11]

Freud was an articulate champion of this idea but not its originator, and the same observation appears in some form or another in the psychological theories of Plato, Aristotle, Hobbes, Mill, Bentham and others. The philosopher and mathematician Blaise Pascal was especially clear on this point:

> All men seek happiness. This is without exception. Whatever different means they employ, they all tend to this end. The cause of some going to war, and of others avoiding it, is

the same desire in both, attended with different views. The will never takes the least step but to this object. This is the motive of every action of every man, even of those who hang themselves.[12]

Feeling Happy Because

If every thinker in every century has recognized that people seek emotional happiness, then how has so much confusion arisen over the meaning of the word? One of the problems is that many people consider the desire for happiness to be a bit like the desire for a bowel movement: something we all have, but not something of which we should be especially proud. The kind of happiness they have in mind is cheap and base – a vacuous state of 'bovine content-ment'[13] that cannot possibly be the basis of a meaningful human life. As the philosopher John Stuart Mill wrote, 'It is better to be a human being dissatisfied than a pig satisfied; better to be Socrates dissatisfied than a fool satisfied. And if the fool, or the pig, are of a different opinion, it is because they only know their own side of the question.'[14]

The philosopher Robert Nozick tried to illustrate the ubiquity of this belief by describing a fictitious virtual-reality machine that would allow anyone to have any experience they chose, and that would conveniently cause them to forget that they were hooked up to the machine.[15] He concluded that no one would willingly choose to get hooked up for the rest of his life because the happiness he would experience with such a machine would not be happiness at all. 'Someone whose emotion is based upon egregiously unjustified and false evaluations we will be reluctant to term happy, however he feels.'[16] In short, emotional happiness is fine for pigs, but it is a goal unworthy of creatures as sophisticated and capable as we.

Now, let's take a moment to think about the difficult position that someone who holds this view is in, and let's guess how they might resolve it. If you considered it perfectly tragic for life to be aimed at nothing more substantive and significant than a *feeling*, and yet you could not help but notice that people spend their days seeking happiness, then what might you be tempted to conclude? Bingo! You might be tempted to conclude that the word *happiness* does not indicate a good feeling but rather that it indicates a very *special* good feeling that can only be produced by very special means – for example, by living one's life in a proper, moral, meaningful, deep, rich, Socratic and non-piglike way. Now *that* would be the kind of feeling one wouldn't be ashamed to strive for. In fact, the Greeks had a word for this kind of happiness – *eudaimonia* – which translates literally as 'good spirit' but which probably means something more like 'human flourishing' or 'life well lived.' For Socrates, Plato, Aristotle, Cicero and even Epicurus (a name usually associated with piggish happiness), the only thing that could induce that kind of happiness was the virtuous performance of one's duties, with the precise meaning of *virtuous* left for each philosopher to work out for himself. The ancient Athenian legislator Solon suggested that one could not say that a person was happy until the person's life had ended because happiness is the result of living up to one's potential – and how can we make such a judgment until we see how the whole thing turns out? A few centuries later, Christian theologians added a nifty twist to this classical conception: happiness was not merely the *product* of a life of virtue but the *reward* for a life of virtue, and that reward was not necessarily to be expected in this lifetime.[17]

For two thousand years philosophers have felt compelled to identify happiness with virtue because that is the sort of happiness they think we *ought* to want. And maybe they're right. But if living one's life virtuously is a cause of happiness, it is not happiness itself, and it does us no good to obfuscate a discussion by calling both the cause and the consequence by the same name. I can produce pain by pricking your finger with a pin or by electrically stimulating a particular spot in your brain, and the two pains will be *identical* feelings produced by different means. It would do us no good to call the first of these *real pain* and the other *fake pain*. Pain is pain, no matter what causes it. By muddling causes and consequences, philosophers have been forced to construct tortured defences of some truly astonishing claims – for example, that a Nazi war criminal who is basking on an Argentinean beach is not really happy, whereas the pious missionary who is being eaten alive by cannibals is. 'Happiness will not tremble,' Cicero wrote in the first century BC, 'however much it is tortured.'[18] That statement may be admired for its courage, but it probably doesn't capture the sentiments of the missionary who was drafted to play the role of the entrée.

Happiness is a word that we generally use to indicate an experience and not the actions that give rise to it. Does it make any sense to say, 'After a day spent killing his parents, Frank was happy'? Indeed it does. We hope there never was such a person, but the sentence is grammatical, well formed and easily understood. Frank is a sick puppy, but if he says he is happy and he looks happy, is there a principled reason to doubt him? Does it make any sense to say, 'Sue was happy to be in a coma'? No, of course not. If Sue is unconscious, she cannot be happy no matter how many good deeds she did before calamity struck. Or how about this one: 'The computer obeyed all Ten Commandments and was happy as a clam'? Again, sorry, but no. There is some remote possibility that clams can be happy because there is some remote possibility that clams have the capacity to feel. There may be something it is like to be a clam, but we can be fairly certain that there is nothing it is like to be a computer, and hence the computer cannot be happy no matter how many of its neighbour's wives it failed to covet.[19] Happiness refers to feelings, virtue refers to actions, and those actions can cause those feelings. But not necessarily and not exclusively.

Feeling Happy About

The you-know-what-I-mean feeling is what people ordinarily mean by *happiness*, but it is not the only thing they mean. If philosophers have muddled the moral and emotional meanings of the word *happiness*, then psychologists have muddled the emotional and judgmental meanings equally well and often. For example, when a person says, 'All in all, I'm happy about the way my life has gone', psychologists are generally willing to grant that the person is happy. The problem is that people sometimes use the word *happy* to express their beliefs about the merits of things, such as when they say, 'I'm happy they caught the little bastard who broke my windshield', and they say things like this even when they are not feeling anything vaguely resembling pleasure. How do we know when a person is expressing a point of view rather than making a claim about her subjective experience? When the word *happy* is followed by the words *that* or *about,* speakers are usually trying to tell us that we ought to take the word *happy* as an indication not of their feelings but rather of their stances. For instance, when our spouse excitedly reveals that she has just been asked to spend six months at the company's new branch in Tahiti while we stay home and mind the kids, we may say, 'I'm not happy, of course, but I'm happy that you're happy.' Sentences such as these make high school English teachers apoplectic, but they are actually quite sensible if we can just resist the temptation to take every instance of the word *happy* as an instance of emotional happiness. Indeed, the first time we utter the word, we are letting our spouse know that we are most certainly not having the you-know-what-I-mean feeling (emotional happiness), and the second time we utter the word we are indicating that we approve of the fact that our spouse is (judgmental happiness). When we say we are happy *about* or happy *that*, we are merely noting that something is a potential source of pleasurable feeling, or a past source of pleasurable feeling, or that we realize it ought to be a source of pleasurable feeling but that it sure doesn't feel that way at the moment. We are not actually claiming to be experiencing the feeling or anything like it. It would be more appropriate for us to tell our spouse, 'I am not happy, but I understand you are, and I can even imagine that were I going to Tahiti and were you remaining home with these juvenile delinquents, I'd be *experiencing* happiness rather than admiring yours.' Of course, speaking like this requires that we forsake all possibility of human companionship, so we opt for the common shorthand and say we are happy *about* things even when we are feeling thoroughly distraught. That's fine, just as long as we keep in mind that we don't always mean what we say.

New Yeller

If we were to agree to reserve the word *happiness* to refer to that class of subjective emotional experiences that are vaguely described as *enjoyable* or *pleasurable*, and if we were to promise not to use that same word to indicate the morality of the actions one might take to induce those experiences or to indicate our judgments about the merits of those experiences, we might still wonder whether the happiness one gets from helping a little old lady across the street constitutes a different kind of emotional experience – bigger, better, deeper – than the happiness one gets from eating a slice of banana-cream pie. Perhaps the happiness one experiences as a result of good deeds *feels* different from that other sort. In fact, while we're at it, we might as well wonder whether the happiness one gets from eating banana-cream pie feels different from the happiness one gets from eating coconut-cream pie. Or from eating a slice of *this* banana-cream pie rather than a slice of *that* one. How can we tell whether subjective emotional experiences are different or the same?

The truth is that we can't – no more than we can tell whether the yellow experience we have when we look at a mustard jar is the same yellow experience that others have when they look at the same mustard jar. Philosophers have flung themselves headlong at this problem for quite some time with little more than bruises to show for it,[20] because when all is said and done, the only way to measure precisely the similarity of two things is for the person who is doing the measuring to compare them side by side – that is, to *experience* them side by side. And outside of science fiction, no one can actually have another person's experience. When we were children, our mothers taught us to call that looking-at-the-mustard-jar experience *yellow*, and being compliant little learners, we did as we were told. We were pleased when it later turned out that everyone else in the kindergarten claimed to experience yellow when they looked at a mustard jar too. But these shared labels may mask the fact that our actual experiences of yellow are quite different, which is why many people do not discover that they are colour-blind until late in life when an ophthalmologist notices that they do not make the distinctions that others seem to make. So while it seems rather unlikely that human beings have radically different experiences when they look at a mustard jar, when they hear a baby cry, or when they smell a former skunk, it *is* possible, and if you want to believe it, then you have every right and no one who values her time should try to reason with you.

Remembering Differences

I hope you aren't giving up *that* easily. Perhaps the way to determine whether a pair of happinesses actually feel different is to forget about comparing the experiences of different minds and just ask someone who has experienced them both. I may never know if *my* experience of yellow is different from *your* experience of yellow, but surely I can tell that my experience of *yellow* is different from my experience of *blue* when I mentally compare the two. Right? Unfortunately, this strategy is more complicated than it looks. The nub of the problem is that when we say that we are mentally comparing two of our own subjective experiences, we are not actually *having* the two experiences at the same time. Rather, we are at best having one of them, having already had the other, and when an interrogator asks us which experience made us happier or whether the two happinesses were the same, we are at best comparing something we are currently experiencing with our *memory* of something we experienced in the past. This would be unobjectionable were it not for the fact that memories – especially memories of experiences – are notoriously unreliable, a fact that has been demonstrated by both magicians and scientists. First the magic. Look at the six royal cards in figure 4, and pick your favourite. No, don't tell me. Keep it to yourself. Just look at your card, and say the name once or twice (or write it down) so that you'll remember it for a few pages.

Fig. 4.

Good. Now consider how scientists have approached the problem of remembered experience. In one study, researchers showed volunteers a colour swatch of the sort one might pick up in the paint aisle of the local hardware store and allowed them to study it for five seconds.[21] Some volunteers then spent thirty seconds describing the colour (describers), while other volunteers did not describe it (non-describers). All volunteers were then shown a lineup of six colour swatches, one of which was the colour they had seen thirty seconds earlier, and were asked to pick out the original swatch. The first interesting finding was that only 73 per cent of the nondescribers were able to identify it accurately. In other words, fewer than three quarters of these people could tell if *this* experience of yellow was the same as the experience of yellow they had had just a half-minute before. The second interesting finding was that describing the colour impaired rather than improved performance on the identification task. Only 33 per cent of the describers were able accurately to identify the original colour. Apparently, the describers' verbal descriptions of their experiences 'overwrote' their memories of the experiences themselves, and they ended up remembering not what they had experienced but what they had *said* about what they experienced. And what they had said was not clear and precise enough to help them recognize it when they saw it again thirty seconds later.

Most of us have been in this position. We tell a friend that we were disappointed with the house chardonnay at that trendy downtown bistro, or with the way the string quartet handled our beloved Bartók's Fourth, but the fact is that we are unlikely to be recalling how the wine actually tasted or how the quartet actually sounded when we make this pronouncement. Rather, we are likely to be recalling that as we left the concert, we mentioned to our companion that both the wine and the music had a promising start and a poor finish. Experiences of chardonnays, string quartets, altruistic deeds and banana-cream pie are rich, complex, multidimensional and impalpable. One of the functions of language is to help us palp them – to help us extract and remember the important features of our experiences so that we can analyse and communicate them later. *The New York Times* online film archive stores critical synopses of films rather than the films themselves, which would take up far too much space, be far too difficult to search and be thoroughly useless to anyone who wanted to know what a film was like without actually seeing it. Experiences are like movies with several added dimensions, and were our brains to store the full-length feature films of our lives rather than their tidy descriptions, our heads would need to be several times larger. And when we wanted to know or tell others whether the tour of the sculpture garden was worth the price of the ticket, we would have to replay the entire episode to find out. Every act of memory would require precisely the amount of time that the event being remembered had originally taken, which would permanently sideline us the first time someone asked if we liked growing up in Chicago. So we reduce our experiences to words such as *happy*, which barely do them justice but which are the things we can carry reliably and conveniently with us into the future. The smell of the rose is unresurrectable, but if we know it was *good* and we know it was *sweet*, then we know to stop and smell the next one.

Perceiving Differences

Our remembrance of things past is imperfect, thus comparing our new happiness with our memory of our old happiness is a risky way to determine whether two subjective experiences are really different. So let's try a slightly modified approach. If we cannot remember the feeling of yesterday's banana-cream pie well enough to compare it with the feeling of today's good deed, perhaps the solution is to compare experiences that are so close together in time that we can actually watch them change. For instance, if we were to do a version of the colour-swatch experiment in which we reduced the amount of time that passed between the presentation of the original swatch and the presentation of the lineup, surely people would have no problem identifying the original swatch, right? So what if we reduced the time to, say, twenty-five seconds? Or fifteen? Ten? How about a fraction of *one*? And what if, as a bonus, we made the identification task a bit easier by showing volunteers a colour swatch for a few seconds, taking it away for just a fraction of a second, and then showing them one test swatch (instead of a lineup of six) and asking them to tell us whether the single test swatch is the same as the original. No intervening verbal description to confuse their memories, no rival test swatches to confuse their eyes, and only a sliver of a slice of a moment between the presentation of the original and test swatches.

Gosh. Given how simple we've made the task, shouldn't we predict that everyone will pass it with, um, flying colours?

Yes, but only if we enjoy being wrong. In a study conceptually similar to the one we just designed, researchers asked volunteers to look at a computer screen and read some odd-looking text.[22] What made the text so odd was that it alternated between uppercase and lowercase, so that it lOoKeD lIkE tHiS. Now, as you may know, when people seem to be staring directly at something, their eyes are actually flickering slightly away from the thing they are staring at three or four times per second, which is why eyeballs look jiggly if you study them up close. The researchers used an eye-tracking device that tells a computer when the volunteer's eyes are fixated on the object on the screen and when they have briefly jiggled away. Whenever the volunteers' eyeballs jiggled away from the text for a fraction of a second, the computer played a trick on them: it changed the case of every letter in the text they were reading so that the text that lOoKeD lIkE tHiS suddenly LoOkEd LiKe ThIs. Amazingly, volunteers did not notice that the text was alternating between different styles several times each second as they read it. Subsequent research has shown that people fail to notice a wide range of these 'visual discontinuities', which is why filmmakers can suddenly change the style of a woman's dress or the colour of a man's hair from one cut to the next, or cause an item on a table to disappear entirely, all without ever waking the audience.[23] Interestingly, when people are asked to predict whether they would notice such visual discontinuities, they are quite confident that they would.[24]

And it isn't just the subtle changes we miss. Even dramatic changes to the appearance of a scene are sometimes overlooked. In an experiment taken straight from the pages of *Candid Camera*, researchers arranged for a researcher to approach pedestrians on a college campus and ask for directions to a particular building.[25] While the pedestrian and the researcher conferred over the researcher's map, two construction workers, each holding one end of a large door, rudely cut between them, temporarily obstructing the pedestrian's view of the researcher. As the construction workers passed, the original researcher crouched down behind the door and walked off with the construction workers, while a new researcher, who had been hiding behind the door all along, took his place and picked up the conversation. The original and substitute researchers were of different heights and builds and had noticeably different voices, haircuts and clothing. You would have no trouble telling them apart if they were standing side by side. So what did the Good Samaritans who had stopped to help a lost tourist make of this switcheroo? Not much. In fact, most of the pedestrians failed to notice – *failed to notice that the person to whom they were talking had suddenly been transformed into an entirely new individual.*

Are we to believe, then, that people cannot tell when their experience of the world has changed right before their eyes? Of course not. If we take this research to its logical extreme we end up as extremists generally do: mired in absurdity and handing out pamphlets. If we could never tell when our experience of the world had changed, how could we know that something was moving, how could we tell whether to stop or go at an intersection, and how could we count beyond one? These experiments tell us that the experiences of our former selves are *sometimes* as opaque to us as the experiences of other people, but more important, they tell us when this is most and least likely to be the case. What was the critical ingredient that allowed each of the foregoing studies to produce the results it did? In each instance, volunteers were not *attending* to their own experience of a particular aspect of a stimulus at the moment of its transition. In the colour-swatch study, the swatches were swapped in another room during the thirty-second break; in the reading study, the text was changed when the volunteer's eye had momentarily jiggled away; in the door study, the researchers switched places only when a large piece of wood was obstructing the volunteer's view. We would not expect these studies to show the same results if burnt umber became fluorescent mauve, or if **this** became **t h a t**, or if an accountant from Poughkeepsie became Queen Elizabeth II while the volunteer was looking right at her, or him, or whatever. And indeed, research has shown that when volunteers are paying close attention to a stimulus at the precise moment that it changes, they do notice that change quickly and reliably.[26] The point of these studies is not that we are hopelessly inept at detecting changes in our experience of the world but rather that unless our minds are keenly focused on a particular aspect of that experience at the very moment it changes, we will be forced to rely on our memories – forced to compare our current experience to our recollection of our former experience – in order to detect the change.

Magicians have known all this for centuries, of course, and have traditionally used their knowledge to spare the rest of us the undue burden of money. A few pages back you chose a card from a group of six. What I didn't tell you at the time was that I have powers far beyond those of mortal men, and therefore I knew which card you were going to pick before you picked it. To prove it, I have removed your card from the group. Take a look at figure 5 and tell me I'm not amazing. How did I do it? This trick is much more exciting, of course, when you don't know beforehand that it's a trick and you don't have to wade through several pages of text to hear the punchline. And it doesn't work at all if you compare the two figures side by side, because you instantly see that none of the cards in figure 4 (including the one you picked) appears in figure 5. But when there is some possibility that the magician knows your chosen card – either by sleight of hand, shrewd deduction or telepathy – and when your jiggly eyes are not looking directly at the first group of six as it transforms into the second group of five, the illusion can be quite powerful. Indeed, when the trick first appeared on a website, some of the smartest scientists I know hypothesized that a newfangled technology was allowing the server to guess their card by tracking the speed and acceleration of their keystrokes. I personally removed my hand from the mouse just to make sure that its subtle movements were not being measured. It did not occur to me until the third time through that while I had *seen* the first group of six cards, I had only *remem-*

Fig. 5.

bered my verbal label for the card I had chosen, and hence had failed to notice that all the other cards had changed as well.[27] What's important to note for our purposes is that card tricks like this work for precisely the same reason that people find it difficult to say how happy they were in their previous marriages.

Happy Talk

Reba and Lori Schappell claim to be happy, and that disturbs us. We are rock-solid certain that it just *can't* be true, and yet, it looks as though there is no foolproof method for comparing their happiness with our own. If they say they are happy, then on what basis can we conclude that they are wrong? Well, we might try the more lawyerly tactic of questioning their ability to know, evaluate or describe their own experience. 'They may *think* they're happy', we could say, 'but that's only because they don't know what happiness really is.' In other words, because Lori and Reba have never had many of the experiences that we singletons have had – spinning cartwheels in a meadow, snorkeling along the Great Barrier Reef, strolling down the avenue without drawing a crowd – we suspect they may have an impoverished background of happy experiences that leads them to evaluate their lives differently than the rest of us would. If, for instance, we were to give the twins a birthday cake, hand them an eight-point rating scale (which can be thought of as an artificial language with eight words for different intensities of happiness), and ask them to report on their subjective experience, they might tell us they felt a joyful *eight*. But isn't it likely that their *eight* and our *eight* represent fundamentally different levels of joy, and that their use of the eight-word language is distorted by their unenviable situation, which has never allowed them to discover how happy a person can really be? Lori and Reba may be using the eight-word language differently than we do because for them, birthday cake is as good as it gets. They label their happiest experience with the happiest word in the eight-word language, naturally, but this should not cause us to overlook the fact that the experience they call *eight* is an experience that we might call *four and a half*. In short, they don't mean *happy* the way we mean *happy*. Figure 6 shows how an impoverished experiential background can cause language to be squished so that the full range of verbal labels is used to describe a restricted range of experiences. By this account, when the twins say they are ecstatic, they are actually *feeling* what we feel when we say we are pleased.

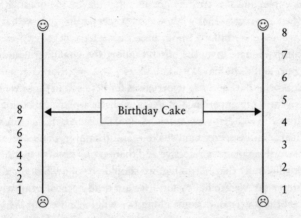

L & R's Experience   Our Experience

Fig. 6. The *language-squishing hypothesis* suggests that when given a birthday cake, Lori and Reba feel exactly as you feel but talk about it differently.

Squishing Language

The nice things about this *language-squishing hypothesis* are *(a)* it suggests that everyone everywhere has the same subjective experience when they receive a birthday cake even if they describe that experience differently, which makes the world a rather simple place to live and bake; and *(b)* it allows us to go on believing that despite what they say about themselves, Lori and Reba aren't *really* happy after all, and thus we are perfectly justified in preferring our lives to theirs. The less nice things about this hypothesis are numerous, and if we worry that Lori and Reba use the eight-word language differently than we do because they have never enjoyed the thrill of a cartwheel, then we had better worry about a few other matters too. For instance, we had better worry that we have never felt the over-whelming sense of peace and security that comes from knowing that a beloved sibling is always by our side, that we will never lose her friendship no matter what kind of crummy stuff we may say or do on a bad day, that there will always be someone who knows us as well as we know ourselves, shares our hopes, worries our worries, and so on. If they haven't had our experiences, then we haven't had theirs either, and it is entirely possible that *we* are the ones with the squished language – that when we say we feel overjoyed, we have no idea what we are talking about because we have never experienced the companionate love, the blissful union, the unadulterated agape that Lori and Reba have. And all of us – you, me, Lori, Reba – had better worry that there are experiences far better than those we have had so far – the experience of flying without a plane, of seeing our children win Academy Awards and Pulitzer Prizes, of meeting God and learning the secret handshake – and that everyone's use of the eight-word language is defective and that *no one* knows what happiness really is. By that reasoning, we should all follow Solon's advice and *never* say we are happy until we are dead because otherwise, if the real thing ever does come along, we will have used up the word and won't have any way to tell the newspapers about it.

But these are just the preliminary worries. There are more. If we wanted to do a thought experiment whose results would demon-strate once and for all that Lori and Reba just don't know what hap-piness really is, perhaps we should imagine that with a wave of a magic wand we could split them apart and allow them to experience life as singletons. If after a few weeks on their own they came to us, repudiated their former claims and begged not to be changed back to their former state, shouldn't that convince us, as it has apparently convinced them, that they were previously confusing their fours and eights? We've all known someone who had a religious conversion, went through a divorce or survived a heart attack and now claims that her eyes are open for the very first time – that despite what she thought and said in her previous incarnation, she was never *really* happy until now. Are the people who have undergone such marvel-lous metamorphoses to be taken at their word?

Not necessarily. Consider a study in which volunteers were

shown some quiz-show questions and asked to estimate the likelihood that they could answer them correctly. Some volunteers were shown only the questions (the question-only group), while others were shown both the questions and the answers (the question-and-answer group). Volunteers in the question-only group thought the questions were quite difficult, while those in the question-and-answer group – who saw both the questions ('What did Philo T. Farnsworth invent?') and the answers ('The television set') – believed that they could have answered the questions easily had they never seen the answers at all. Apparently, once volunteers knew the answers, the questions seemed simple ('Of course it was the television – everyone knows that!'), and the volunteers were no longer able to judge how difficult the questions would seem to someone who did not share their knowledge of the answers.[28]

Studies such as these demonstrate that once we have an experience, we cannot simply set it aside and see the world as we would have seen it had the experience never happened. To the judge's dismay, the jury cannot disregard the prosecutor's snide remarks. Our experiences instantly become part of the lens through which we view our entire past, present and future, and like any lens, they shape and distort what we see. This lens is not like a pair of spectacles that we can set on the nightstand when we find it convenient to do so but like a pair of contacts that are forever affixed to our eyeballs with superglue. Once we learn to read, we can never again see letters as mere inky squiggles. Once we learn about free jazz, we can never again hear Ornette Coleman's saxophone as a source of noise. Once we learn that van Gogh was a mental patient, or that Ezra Pound was an anti-Semite, we can never again view their art in the same way. If Lori and Reba were separated for a few weeks, and if they told us that they were happier now than they used to be, they might be right. But they might not. They might just be telling us that the singletons they had become now viewed being conjoined with as much distress as those of us who have always been singletons do. Even if they could remember what they thought, said and did as conjoined twins, we would expect their more recent experience as singletons to colour their evaluation of the conjoined experience, leaving them unable to say with certainty how conjoined twins who had never been singletons actually feel. In a sense, the experience of separation would make them *us,* and thus they would be in the same difficult position that we are in when we try to imagine the experience of being conjoined. Becoming singletons would affect their views of the past in ways that they could not simply set aside. All of this means that when people have new experiences that lead them to claim that their language was squished – that they were not really happy even though they said so and thought so at the time – they can be mistaken. In other words, people can be wrong in the present when they say they were wrong in the past.

Stretching Experience

Lori and Reba have not done many of the things that for the rest of us give rise to feelings near the top of the happiness scale – cartwheels, scuba diving, name your poison – and surely this must make a difference. If impoverished experiential backgrounds don't necessarily squish language, then what do they do instead? Let's assume that Lori and Reba really do have an impoverished experiential background against which to evaluate something as simple as, say, the dutiful presentation of a chocolate cake on their birthday. One possibility is that their impoverished experiential background would squish their language. But another possibility is that their impoverished experiential background would not squish their language so much as it would stretch their experience – that is, when they say *eight* they mean exactly the same thing we mean when we say *eight* because when they receive a birthday cake they feel exactly the same way that the rest of us feel when we do underwater cartwheels along the Great Barrier Reef. Figure 7 illustrates the *experience-stretching hypothesis.*

Experience stretching is a bizarre phrase but not a bizarre idea. We often say of others who claim to be happy despite circumstances that we believe should preclude it that 'they only think they're happy because they don't know what they're missing'. Okay, sure, *but that's the point.* Not knowing what we're missing can mean that we are truly happy under circumstances that would not allow us to be happy once we have experienced the missing thing. It does *not* mean that those who don't know what they're missing are *less* happy than

L & R's Experience   Our Experience

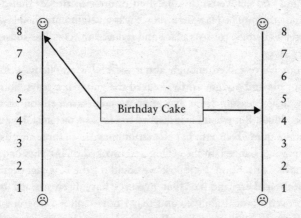

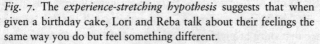

Fig. 7. The *experience-stretching hypothesis* suggests that when given a birthday cake, Lori and Reba talk about their feelings the same way you do but feel something different.

those who have it. Examples abound in my life and yours, so let's talk about mine. I occasionally smoke a cigar because it makes me happy, and my wife occasionally fails to understand why I must have a cigar to be happy when she can apparently be just as happy without one (and even happier without me having one). But the experience-stretching hypothesis suggests that I too could have been happy without cigars if only I had not experienced their pharmacological mysteries in my wayward youth. But I did, and because I did I now know what I am missing when I don't, hence that glorious moment during my spring holiday when I am reclining in a lawn chair on the golden sands of Kauai, sipping Talisker and watching the sun slip slowly into a taffeta sea, is just not quite perfect if I don't also have something stinky and Cuban in my mouth. I could press both my luck and my marriage by advancing the language-squishing hypothesis, carefully explaining to my wife that because she has never experienced the pungent earthiness of a Montecristo no. 4, she has an impoverished experiential background and therefore does not know what happiness really is. I would lose, of course, because I always do, but in this case I would deserve it. Doesn't it make better sense to say that by learning to enjoy cigars I changed my experiential back-ground and inadvertently ruined all future experiences that do not include them? The Hawaiian sunset was an eight until the Hawaiian sunset à la stogie took its place and reduced the cigarless sunset to a mere seven.[29]

But we've talked enough about me and my holiday. Let's talk about me and my guitar. I've played the guitar for years, and I get very little pleasure from executing an endless repetition of three-chord blues. But when I first learned to play as a teenager, I would sit upstairs in my bedroom happily strumming those three chords until my parents banged on the ceiling and invoked their rights under the Geneva Convention. I suppose we could try the language-squishing hypothesis here and say that my eyes have been opened by my improved musical abilities and that I now realize I was not *really* happy in those teenage days. But doesn't it seem more reasonable to invoke the experience-stretching hypothesis and say that an experi-ence that once brought me pleasure no longer does? A man who is given a drink of water after being lost in the Mojave Desert for a week may at that moment rate his happiness as eight. A year later, the same drink might induce him to feel no better than two. Are we to believe that he was wrong about how happy he was when he took that life-giving sip from a rusty canteen, or is it more reasonable to say that a sip of water can be a source of ecstasy or a source of mois-ture depending on one's experiential background? If impoverished experiential backgrounds squish our language rather than stretch our experience, then children who say they are delighted by peanut butter and jelly are just plain wrong, and they will admit it later in life when they get their first bite of goose liver, at which time they will be right, until they get older and begin to get heartburn from fatty foods, at which time they will realize that they were wrong then too. Every day would be a repudiation of the day before, as we experienced greater and greater happiness and realized how thor-oughly deluded we were until, conveniently enough, now.

So which hypothesis is correct? We can't say. What we *can* say is that all claims of happiness are claims from someone's *point of view* – from the perspective of a single human being whose unique collec-tion of past experiences serves as a context, a lens, a background for her evaluation of her current experience. As much as the scientist might wish for it, there isn't a view from nowhere. Once we have an experience, we are thereafter unable to see the world as we did before. Our innocence is lost and we cannot go home again. We may remember what we thought or said (though not necessarily), and we may remember what we did (though not necessarily that either), but the likelihood is depressingly slim that we can resurrect our experience and then evaluate it as we would have back then. In some ways, the cigar-smoking, guitar-playing, pâté-eating people we become have no more authority to speak on behalf of the people we used to be than do outside observers. The separated twins may be able to tell us how they *now* feel about having been conjoined, but they cannot tell us how conjoined twins who have never experienced separation feel about it. No one knows if Reba's and Lori's eight feels like our eight, and that includes all the Rebas and Loris that will ever be.

Onward

On the morning of 15 May 1916, the arctic explorer Ernest Shackleton began the last leg of one of history's most gruelling adventures. His ship, the *Endurance*, had sunk in the Weddell Sea, stranding him and his crew on Elephant Island. After seven months, Shackleton and five of his crewmen boarded a small lifeboat in which they spent three weeks crossing eight hundred miles of frigid, raging ocean. Upon reaching South Georgia Island, the starving, frostbitten men prepared to disembark and cross the island on foot in the hope of reaching a whaling station on the other side. No one had ever survived that trek. Facing almost certain death that morning, Shackleton wrote:

> We passed through the narrow mouth of the cove with the ugly rocks and waving kelp close on either side, turned to the east, and sailed merrily up the bay as the sun broke through the mists and made the tossing waters sparkle around us. We were a curious-looking party on that bright morning, but we were feeling happy. We even broke into song, and, but for our

Robinson Crusoe appearance, a casual observer might have taken us for a picnic party sailing in a Norwegian fjord or one of the beautiful sounds of the west coast of New Zealand.[30]

Could Shackleton really have meant what he said? Could his *happy* be our *happy*, and is there any way to tell? As we've seen, happiness is a subjective experience that is difficult to describe to ourselves and to others, thus evaluating people's claims about their own happiness is an exceptionally thorny business. But don't worry – because before business gets better, it gets a whole lot thornier.

CHAPTER 3

Outside Looking In

Go to your bosom;
Knock there, and ask your heart what it doth know.

Shakespeare, *Measure for Measure*

THERE AREN'T MANY JOKES about psychology professors, so we tend to cherish the few we have. Here's one. What do psychology professors say when they pass each other in the hallway? 'Hi, you're fine, how am I?' I know, I know. The joke isn't that funny. But the reason it's *supposed* to be funny is that people shouldn't know how others are feeling but they should know how they're feeling themselves. 'How are you?' is overly familiar for the same reason that 'How am I?' is overly strange. And yet, strange as it is, there are times when people seem not to know their own hearts. When conjoined twins claim to be happy, we have to wonder if perhaps they just *think* they're happy. That is, they may believe what they're saying, but what they're saying may be wrong. Before we can decide whether to accept people's claims about their happiness, we must first decide whether people can, in principle, be mistaken about what they feel. We can be wrong about all sorts of things – the price of soybeans, the life span of dust mites, the history of flannel – but can we be wrong about our own emotional experience? Can we believe we are feeling something we aren't? Are there really people out there who can't accurately answer the world's most familiar question?

Yes, and you'll find one in the mirror. Read on.

Dazed and Confused

But not just yet. Before you read on, I challenge you to stop and have a nice long look at your thumb. Now, I will wager that you did not accept my challenge. I will wager that you went right on reading because looking at your thumb is so easy that it makes for rather pointless sport – everyone bats a thousand and the game is called on account of boredom. But if looking at your thumb seems beneath you, just consider what actually has to happen for us to see an object in our environment – a thumb, a glazed doughnut or a rabid wolverine. In the tiny gap between the time that the light reflected from the surface of the object reaches our eyes and the time that we become aware of the object's identity, our brains must extract and analyse the object's features and compare them with information in our memory to determine what the thing is and what we ought to do about it. This is complicated stuff – so complicated that no scientist yet understands precisely how it happens and no computer can simulate the trick – but it is just the sort of complicated stuff that brains do with exceptional speed and accuracy. In fact, they perform these analyses with such proficiency that we have the experience of simply looking leftward, seeing a wolverine, feeling afraid and preparing to do all further analysis from the safety of a sycamore tree.

Think for a moment about how looking *ought* to happen. If you were designing a brain from scratch, you would probably design it so that it *first* identified objects in its environment ('Sharp teeth, brown fur, weird little snorting sound, hot drool – why, that's a rabid wolverine!') and *then* figured out what to do ('Leaving seems like a splendid idea about now'). But human brains were not designed from scratch. Rather, their most critical functions were designed first, and their less critical functions were added on like bells and whistles as the millennia passed, which is why the really important parts of your brain (e.g., the ones that control your breathing) are down at the bottom and the parts you could probably live without (e.g., the ones that control your temper) sit above them, like ice cream on a cone. As it turns out, running with great haste from rabid wolverines is much more important than knowing what they are. Indeed, actions such as running away are so vitally important to the survival of terrestrial mammals like the ones from whom we are descended that evolution took no chances and designed the brain to answer the 'What should I do?' question *before* the 'What is it?' question.[1] Experiments have demonstrated that the moment we encounter an object, our brains instantly analyse just a few of its key features and then use the presence or absence of these features to make one very fast and very simple decision: 'Is this object an important thing to which I ought to respond right now?'[2] Rabid wolverines, crying babies, hurled rocks, beckoning mates, cowering prey – these things count for a lot in the game of survival, which requires that we take immediate action when we happen upon them and do not dally to contemplate the finer points of their identities. As such, our brains are designed to decide *first* whether objects count and to decide later what those objects are. This means that when you turn your head to the left, there is a fraction of a second during which your brain does *not* know that it is seeing a wolverine but *does* know that it is seeing something scary.

But how can that be? How can we know something is scary if we don't know what it is? To understand how this can happen, just consider how you would go about identifying a person who is walking toward you across a vast expanse of desert. The first thing to catch your eye would be a small flicker of motion on the horizon. As you stared, you would soon notice that the motion was that of an object moving toward you. As it came closer, you would see that the motion was biological, then you would see that the biological object was a biped, then a human, then a female, then a fat human female with dark hair and a Budweiser T-shirt, and then – hey, what's Aunt Mabel doing in the Sahara? Your identification of Aunt Mabel would *progress* – that is, it would begin quite generally and become more specific over time, until finally it terminated in a family reunion. Similarly, the identification of a wolverine at your elbow progresses over time – albeit just a few milliseconds – and it too progresses from the general to the specific. Research demonstrates that there is enough information in the early, very general stages of this identification process to decide whether an object is scary, but not enough information to know what the object is. Once our brains decide that they are in the presence of a threat, they instruct our glands to produce hormones that create a state of heightened physiological arousal – blood pressure rises, heart rate increases, pupils dilate, muscles tense – which prepares us to spring into action. Before our brains have finished the full-scale analysis that will allow us to know that the object is a wolverine, they have already put our bodies into their ready-to-run-away modes – all pumped up and raring to go.

The fact that we can feel aroused without knowing exactly what it is that has aroused us has important implications for our ability to identify our own emotions.[3] For example, researchers studied the reactions of some young men who were crossing a long, narrow suspension bridge constructed of wooden boards and wire cables that rocked and swayed 230 feet above the Capilano River in North Vancouver.[4] A young woman approached each man and asked if he would mind completing a survey, and after he did so, the woman gave the man her telephone number and offered to explain her survey project in greater detail if he called. Now, here's the catch: the woman approached some of these young men as they were crossing the bridge and others only after they had crossed it. As it turned out, the men who had met the woman as they were crossing the bridge were much more likely to call her in the coming days. Why? The men who met the woman in the middle of a shaky, swaying suspension bridge were experiencing intense physiological arousal, which they would normally have identified as fear. But because they were being interviewed by an attractive woman, they mistakenly identified their arousal as sexual attraction. Apparently, feelings that one interprets as fear in the presence of a sheer drop may be interpreted as lust in the presence of a sheer blouse – which is simply to say that people *can* be wrong about what they are feeling.[5]

Comfortably Numb

The novelist Graham Greene wrote: 'Hatred seems to operate the same glands as love.'[6] Indeed, research shows that physiological arousal can be interpreted in a variety of ways, and our interpretation of our arousal depends on what we believe caused it. It is possible to mistake fear for lust, apprehension for guilt,[7] shame for anxiety.[8] But just because we don't always know what to *call* our emotional experience doesn't mean that we don't know what that experience is *like*, does it? Perhaps we can't say its name and perhaps we don't know what made it happen, but we always know what it feels like, right? Is it possible to believe we are feeling *something* when we are actually feeling *nothing at all*? The philosopher Daniel Dennett put the question this way:

> Suppose someone is given the post-hypnotic suggestion that upon awakening he will *have* a pain in his wrist. If the hypnosis works, is it a case of pain, hypnotically induced, or merely a case of a person who has been induced to *believe* he has a pain? If one answers that the hypnosis has induced real pain, suppose the post-hypnotic suggestion had been: 'On awakening you will *believe* you have a pain in the wrist.' If this suggestion works, is the circumstance just like the previous one? Isn't believing you are in pain tantamount to being in pain?[9]

At first blush, the idea that we can mistakenly believe we are feeling pain seems preposterous, if only because the distinction between *feeling pain* and *believing one is feeling pain* looks so suspiciously like an artefact of language. But give this idea a second blush while considering the following scenario. You are sitting at a pavement café, sipping a tangy espresso and contentedly browsing the Sunday newspaper. People are strolling by and taking in the fine morning, and the amorous activities of a young couple at a nearby table attest to the eternal wonder of spring. The song of a thrush punctuates the yeasty scent of new croissants that wafts from the bakery. The article you are reading on campaign-finance reform is quite interesting and all is well – until suddenly you realize you are now reading the third paragraph, that somewhere in the middle of the first you started sniffing baked goods and listening to bird chirps, and that you now have absolutely no idea what the story you are reading is about. Did you actually read that second paragraph, or did you merely dream it? You take a quick look back and, sure enough, all the words are familiar. As you read them again you can even recall hearing them spoken a few moments ago by that narrator in your head who sounds astonishingly like you and whose voice was submerged for a paragraph or two beneath the sweet distractions of the season.

Two questions confront us. First, did you experience the paragraph the first time you read it? Second, if so, did you know you were experiencing it? The answers are yes and no, respectively. You experienced the paragraph and that's why it was so familiar to you when you went back through it. Had there been an eye tracker at your table, it would have revealed that you did not stop reading at any point. In fact, you were smack-dab in the middle of reading's smooth movements when suddenly you caught yourself . . . caught yourself . . . caught yourself *what*? Experiencing without being aware that you were experiencing – that's what. Now, let me slow down for a moment and tread carefully around these words lest you start listening for the high-pitched tones of the indigo bunting. The word *experience* comes from the Latin *experientia*, meaning 'to try', whereas the word *aware* comes from the Greek *horan*, meaning 'to see'. Experience implies participation in an event, whereas awareness implies observation of an event. The two words can normally be substituted in ordinary conversation without much damage, but they are differently inflected. One gives us the sense of being engaged, whereas the other gives us the sense of being cognizant of that engagement. One denotes reflection while the other denotes the thing being reflected. In fact, awareness can be thought of as a kind of experience of our own experience.[10] When two people argue about whether their dogs are conscious, one is usually using that badly bruised term to mean 'capable of experience' while the other is using it to mean 'capable of awareness'. Dogs are not rocks, one argues, so of course they are conscious. Dogs are not people, the other replies, so of course they are not conscious. Both arguers are probably right. Dogs probably do have an experience of yellow and sweet: there is something it is like to be a dog standing before a sweet, yellow thing, even if human beings can never know what that something is. But the experiencing dog is probably not simultaneously aware that it is having that experience, thinking as it chews, 'Damned fine ladyfinger'.

The distinction between experience and awareness is elusive because most of the time they hang together so nicely. We pop a ladyfinger into our mouths, we experience sweetness, we know we are experiencing sweetness, and nothing about any of this seems even remotely challenging. But if the typically tight bond between experience and awareness leads us to suspect that the distinction between them is an exercise in hand waving, you need only rewind the tape a bit and imagine yourself back at the café at precisely the moment that your eyes were running across the newsprint and your mind was about to mosey off to contemplate the sounds and smells around you. Now hit *play* and imagine that your mind wanders away, gets lost and never comes back. That's right. Imagine that as you experience the newspaper article, your awareness becomes permanently unbound from your experience, and you never catch yourself drifting away – never return to the moment with a start to discover that you are reading. The young couple at the nearby table stop pawing each other long enough to lean over and ask you for the latest news on the campaign-finance reform bill, and you patiently explain that you could not possibly know that because, as they would surely see if only they would pay attention to something other than their glands, you are happily listening to the sounds of spring and *not reading a newspaper*. The young couple is perplexed by this response, because as far as they can see, you do indeed have a newspaper in your hands and your eyeballs are, in fact, running rapidly across the page even as you deny it. After a bit of whispering and one more smooch, they decide to run a test to determine whether you are telling the truth. 'Sorry to bother you again, but we are desperate to know how many senators voted for the campaign-finance reform bill last week and wonder if you would be good enough to hazard a guess?' Because you are sniffing croissants, listening to bird calls and *not reading a newspaper*, you have no idea how many senators voted for the bill. But it appears that the only way to get these strange people to mind their own business is to tell them *something*, so you pull a number out of thin air. 'How about forty-one?' you offer. And to no one's astonishment but your own, the number is exactly right.

This scenario may seem too bizarre to be real (after all, how likely is it that forty-one senators would actually vote for campaign-finance reform?), but it is both. Our visual experience and our awareness of that experience are generated by different parts of our brains, and as such, certain kinds of brain damage (specifically, lesions to the primary visual cortical receiving area known as V1) can impair one without impairing the other, causing experience and awareness to lose their normally tight grip on each other. For example, people who suffer from the condition known as *blindsight* have no awareness of seeing, and will truthfully tell you that they are completely blind.[11] Brain scans lend credence to their claims by revealing diminished activity in the areas normally associated with awareness of visual experience. On the other hand, the same scans reveal relatively normal activity in the areas associated with vision.[12] So if we flash a light on a particular spot on the wall and ask the blindsighted person if she saw the light we just flashed, she tells us, 'No, of course not. As you might infer from the presence of the guide dog, I'm blind.' But if we ask her to make a guess about where the light might have appeared – *just take a stab at it, say anything, point randomly if you like* – she 'guesses' correctly far more often than we would expect by chance. She is *seeing*, if by *seeing* we mean experiencing the light and acquiring knowledge about its location, but she is *blind*, if by *blind* we mean that she is not aware of having seen. Her eyes are projecting the movie of reality on the little theatre screen in her head, but the audience is in the lobby getting popcorn.

This dissociation between awareness and experience can cause the same sort of spookiness with regard to our emotions. Some people seem to be keenly aware of their moods and feelings, and may even have a novelist's gift for describing their every shade and flavour. Others of us come equipped with a somewhat more basic emotional vocabulary that, much to the chagrin of our romantic partners, consists primarily of *good*, *not so good* and *I already told you*. If our expressive deficit is so profound and protracted that it even occurs outside of football season, we may be diagnosed with *alexithymia*, which literally means 'absence of words to describe emotional states'. When alexithymics are asked *what* they are feeling, they usually say, 'Nothing', and when they are asked *how* they are feeling, they usually say, 'I don't know.' Alas, theirs is not a malady that can be cured by a pocket thesaurus or a short course in word power, because alexithymics do not lack the traditional affective lexicon so much as they lack introspective awareness of their emotional states. They seem to *have* feelings, they just don't seem to know about them. For instance, when researchers show volunteers emotionally evocative pictures of amputations and car wrecks, the physiological responses of alexithymics are indistinguishable from those of normal people. But when they are asked to make verbal ratings of the unpleasantness of those pictures, alexithymics are decidedly less capable than normal people of distinguishing them from pictures of rainbows and puppies.[13] Some evidence suggests that alexithymia is caused by a dysfunction of the anterior cingulate cortex, which is a part of the brain known to mediate our awareness of many things, including our inner states.[14] Just as the decoupling of awareness and visual experience can give rise to blindsight, so the decoupling of awareness and emotional experience can give rise to what we might call *numbfeel*. Apparently, it *is* possible – at least for some of the people some of the time – to be happy, sad, bored or curious and not know it.

Warm the Happyometer

Once upon a time there was a bearded God who made a small, flat earth, and pasted it in the very middle of the sky so that human beings would be at the centre of everything. Then physics came along and complicated the picture with big bangs, quarks, branes and superstrings, and the payoff for all that critical analysis is that now, several hundred years later, most people have no idea where they are. Psychology has also created problems where once there were none by exposing the flaws in our intuitive understandings of ourselves. Maybe the universe has several small dimensions tucked inside the large ones, maybe time will eventually stand still or flow backward, and maybe people like us were never meant to fathom a bit of it. But one thing we can always count on is our own experience. The philosopher and mathematician René Descartes concluded that our experience is the *only* thing about which we may be completely sure and that everything else we think we know is merely an inference from that. And yet, we have seen that when we say with moderate precision what we mean by words such as *happiness*, we still can't be sure that two people who claim to be happy are having the same experience, or that our current experience of happiness is really different from our past experience of happiness, or that we are *having* an experience of happiness at all. If the goal of science is to make us feel awkward and ignorant in the presence of things we once understood perfectly well, then psychology has succeeded above all others.

But like happiness, *science* is one of those words that means too many things to too many people and is thus often at risk of meaning nothing at all. My father is an eminent biologist who, after pondering the matter for some decades, recently revealed to me that psychology can't really be a science because science requires the use of electricity. Apparently shocks to your ankles don't count. My own definition of science is a bit more eclectic, but one thing about which I, my dad and most other scientists can agree is that if a thing cannot be measured, then it cannot be studied scientifically. It can be studied, and one might even argue that the study of such unquantifiables is more worthwhile than all the sciences laid end to end. But it is not science because science is about measurement, and if a thing cannot be measured – cannot be compared with a clock or a ruler or something other than itself – it is not a potential object of scientific enquiry. As we have seen, it is extremely difficult to measure an individual's happiness and feel completely confident in the validity and reliability of that measurement. People may not know how they feel, or remember how they felt, and even if they do, scientists can never know exactly how their experience maps onto their description of that experience, and hence they cannot know precisely how to interpret people's claims. All of this suggests that the scientific study of subjective experience is bound to be tough going.

Tough, yes, but not impossible, because the chasm between experiences can be bridged – not with steel girders or a six-lane toll road, mind you, but with a length of reasonably sturdy rope – if we accept three premises.

Measuring Right

The first premise is something that any carpenter could tell you: imperfect tools are a real pain, but they sure beat pounding nails with your teeth. The nature of subjective experience suggests there will never be a *happyometer* – a perfectly reliable instrument that allows an observer to measure with complete accuracy the characteristics of another person's subjective experience so that the measurement can be taken, recorded and compared with another.[15] If we demand that level of perfection from our tools, then we better pack up the eye trackers, brain scanners and colour swatches and cede the study of subjective experience to the poets, who did a nice job with it for the first few thousand years. But if we do that, then it is only fair that we hand them the study of almost everything else as well. Chronometers, thermometers, barometers, spectrometers and every other device that scientists use to measure the objects of their interest are imperfect. Every one of them introduces some degree of error into the observations it allows, which is why governments and universities pay obscene sums of money each year for the slightly more perfect version of each. And if we are purging ourselves of all things that afford us only imperfect approximations of the truth, then we need to discard not only psychology and the physical sciences but law, economics and history as well. In short, if we adhere to the standard of perfection in all our endeavours, we are left with nothing but mathematics and the White Album. So maybe we just need to accept a bit of fuzziness and stop complaining.

The second premise is that of all the flawed measures of subjective experience that we can take, the honest, real-time report of the attentive individual is the least flawed.[16] There are many other ways to measure happiness, of course, and some of them *appear* to be much more rigorous, scientific and objective than a person's own claims. For example, electromyography allows us to measure the electrical signals produced by the striated muscles of the face, such as the *corrugator supercillia*, which furrows our brows when we experience something unpleasant, or the *zygomaticus major*, which pulls our mouths up toward our ears when we smile. Physiography allows us to measure the electrodermal, respiratory and cardiac activity of the autonomic nervous system, all of which change when we experience strong emotions. Electroencephalography, positron-emission tomography and magnetic resonance imaging allow us to measure electrical activity and blood flow in different regions of the brain, such as the left and right prefrontal cortex, which tend to be active when we are experiencing positive and negative emotions, respectively. Even a clock can be a useful device for measuring happiness, because startled people tend to blink more slowly when they are feeling happy than when they are feeling fearful or anxious.[17]

Scientists who rely on the honest, real-time reports of attentive individuals often feel the need to defend that choice by reminding us that these reports correlate strongly with other measures of happiness. But in a sense, they've got it backward. After all, the only reason why we take any of these bodily events – from muscle movement to cerebral blood flow – as indices of happiness is that *people tell us they are*. If everyone claimed to feel raging anger or thick, black depression when their zygomatic muscle contracted, their eyeblink slowed and the left anterior brain region filled with blood, then we would have to revise our interpretations of these physiological changes and take them as indices of unhappiness instead. If we want to know how a person feels, we must begin by acknowledging the fact that there is one and only one observer stationed at the critical point of view. She may not always remember what she felt before, and she may not always be aware of what she is feeling right now. We may be puzzled by her reports, sceptical of her memory, and worried about her ability to use language as we do. But when all our hand wringing is over, we must admit that she is the *only* person who has even the *slightest* chance of describing 'the view from in here', which is why her claims serve as the gold standard against which all other measures are measured. We will have greater confidence in her claims when they jibe with what other, less privileged observers tell us, when we feel confident that she evaluates her experience against the same background that we do, when her body does what most other bodies do when they experience what she is claiming to experience, and so on. But even when all of these various indices of happiness dovetail nicely, we cannot be perfectly sure that we know the truth about her inner world. We can, however, be sure that we have come as close as observers ever get, and that *has* to be good enough.

Measuring Often

The third premise is that imperfections in measurement are always a problem, but they are a devastating problem only when we don't recognize them. If we have a deep scratch on our eyeglasses and don't know it, we may erroneously conclude that a small crack has opened in the fabric of space and is following us wherever we go. But if we are cognizant of the scratch, we can do our best to factor it out of our observations, reminding ourselves that what looks like a rip in space is really just a flaw in the device we are using to observe it. What can scientists do to 'see through' the flaws inherent in reports of subjective experiences? The answer lies in a phenomenon that statisticians call the *law of large numbers*.

Many of us have a mistaken idea about large numbers, namely, that they are like small numbers, only bigger. As such, we expect them to do *more* of what small numbers do but not to do anything *different*. So, for instance, we know that two neurons swapping electrochemical signals across their axons and dendrites cannot possibly be conscious. Nerve cells are simple devices, less complex than walkie-talkies from Sears, and they do one simple thing, namely, react to the chemicals that reach them by releasing chemicals of their own. If we blithely go on to assume that ten billion of these simple devices can only do ten billion simple things, we would never guess that billions of them can exhibit a property that two, ten or ten thousand cannot. Consciousness is precisely this sort of *emergent property* – a phenomenon that arises in part as a result of the sheer *number* of interconnections among neurons in the human brain and that does not exist in any of the parts or in the interconnection of just a few.[18] Quantum physics offers a similar lesson. We know that subatomic particles have the strange and charming ability to exist in two places at once, and if we assume that anything composed of these particles must behave likewise, we should expect all cows to be in all possible barns at the same time. Which they obviously are not, because fixedness is another one of those properties that emerges from the interaction of a terribly large number of terribly tiny parts that do not themselves have it. In short, more is not just more – it is sometimes *other* – than less.

The magic of large numbers works along with the laws of probability to remedy many of the problems associated with the imperfect measurement of subjective experience. You know that if a fair coin is flipped on several occasions it should come up heads about half the time. As such, if you have nothing better to do on a Tuesday evening, I invite you to meet me at the Grafton Street Pub in Harvard Square and play an endearingly mindless game called Splitting the Tab with Dan. Here's how it works. We flip a coin, I call heads, you call tails, and the loser pays the good barkeep, Paul, for our beers each time. Now, if we flipped the coin four times and I won on three of them, you would undoubtedly chalk it up to bad luck on your part and challenge me to darts. But if we flipped the coin four million times and I won on three million of them, then you and your associates would probably send out for a large order of tar and feathers. Why? Because even if you don't know the first thing about probability theory, you have a very keen intuition that when numbers are small, little imperfections – like a stray gust of wind, or a dab of perspiration on a finger – can influence the outcome of a coin flip. But when numbers are large, such imperfections stop mattering. There may have been a dollop of sweat on the coin on a few of the flips, and there may have been a wayward puff of air on a few others, and these imperfections might well account for the fact that the coin came up heads once more than expected when we flipped it four times. But what are the odds that these imperfections could have caused the coin to come up heads a million more times than expected? Infinitesimal, your intuition tells you, and your intuition is spot on. The odds are as close to infinitesimal as things on earth get without disappearing altogether.

This same logic can be applied to the problem of subjective experience. Suppose we were to give a pair of volunteers a pair of experiences that were meant to induce happiness – say, by giving a million dollars to one of them and the gift of a small-calibre revolver to the other. We then ask each volunteer to tell us how happy he or she is. The nouveau riche volunteer says she is ecstatic, and the armed volunteer says he is mildly pleased (though perhaps not quite as pleased as one ought to make an armed volunteer). Is it possible that the two are actually having the same subjective emotional experiences but describing them differently? Yes. The new millionaire may be demonstrating politeness rather than joy. Or perhaps the new pistol owner is experiencing ecstasy but, because he recently shook the hand of God near the Great Barrier Reef, is describing his ecstasy as mere satisfaction. These problems are real problems, significant problems, and we would be foolish to conclude on the basis of these two reports that happiness is not, as it were, a warm gun. But if we gave away *a million pistols* and *a million envelopes of money*, and if 90 percent of the people who got new money claimed to be happier than 90 percent of the people who got new weapons, the odds that we are being deceived by the idiosyncrasies of verbal descriptions become very small indeed. Similarly, if a person tells us that she is happier with today's banana-cream pie than with yesterday's coconut-cream pie, we may rightfully worry that she is misremembering her prior experience. But if this were to happen over and over again with hundreds or thousands of people, some of whom tasted the coconut-cream pie before the banana-cream pie and some of whom tasted it after, we would have good reason to suspect that different pies really do give rise to different experiences, one of which is more pleasant than the other. After all, what are the odds that *everyone* misremembers banana-cream pie as better and coconut-cream pie as worse than they really were?

The fundamental problem in the science of experience is that if either the language-squishing hypothesis or the experience-stretching hypothesis is correct, then every one of us may have a different mapping of what we experience onto what we say – and because subjective experiences can be shared only by saying, the true nature of those experiences can never be perfectly measured. In other words, if the experience and description scales are calibrated a bit differently for every person who uses them, then *it is impossible for scientists to compare the claims of two people.* That's a problem. But the problem isn't with the word *compare*, it's with the word *two*. Two is too small a number, and when it becomes two hundred or two thousand, the different calibrations of different individuals begin to cancel one another out. If the workers at the factory that makes all the world's tape measures, rulers and yardsticks got sloshed at a holiday party and started turning out millions of slightly different-sized measuring instruments, we would not feel confident that a dinosaur was larger than a turnip if you measured one and I measured the other. After all, we may have used pickled rulers. But if hundreds of people with hundreds of rulers stepped up to one of these objects and took its measurements, we could average those measurements and feel reasonably confident that a tyrannosaurus is indeed bigger than a root vegetable. After all, what are the odds that all the people who measured the dinosaur just so happened to have used stretched rulers, and that all the people who measured the turnip just so happened to have used squished rulers? Yes, it is *possible*, and the odds can be calculated quite precisely, but I will spare you the math and promise you that they are so slender that writing them down would endanger the world's supply of zeroes.

The bottom line is this: the attentive person's honest, real-time report is an imperfect approximation of her subjective experience, but it is the only game in town. When a fruit salad, a lover or a jazz trio is just too imperfect for our tastes, we stop eating, kissing and listening. But the law of large numbers suggests that when a measurement is too imperfect for our tastes, we should not stop measuring. Quite the opposite – we should measure again and again until niggling imperfections yield to the onslaught of data. Those subatomic particles that like to be everywhere at once seem to cancel out one another's behavior so that the large conglomeration of particles that we call cows, cars and French Canadians stay exactly where we put them. By the same logic, the careful collection of a large number of experiential reports allows the imperfections of one to cancel out the imperfections of another. No individual's report may be taken as an unimpeachable and perfectly calibrated index of his experience – not yours, not mine – but we can be confident that if we ask enough people the same question, the average answer will be a roughly accurate index of the average experience. The science of happiness requires that we play the odds, and thus the information it provides us is always at some risk of being wrong. But if you want to bet against it, then flip that coin one more time, get out your wallet and tell Paul to make mine a Guinness.

Onward

One of the most annoying songs in the often annoying history of popular music begins with this line: 'Feelings, nothing more than feelings.' I wince when I hear it because it always strikes me as roughly equivalent to starting a hymn with 'Jesus, nothing more than Jesus.' Nothing *more* than feelings? What could be more important than feelings? Sure, *war* and *peace* may come to mind, but are war and peace important for any reason other than the feelings they produce? If war didn't cause pain and anguish, if peace didn't provide for delights both transcendental and carnal, would either of them matter to us at all? War, peace, art, money, marriage, birth, death, disease, religion – these are just a few of the Really Big Topics over which oceans of blood and ink have been spilled, but they are really big topics for one reason alone: each is a powerful source of human emotion. If they didn't make us *feel* uplifted, desperate, thankful and hopeless, we would keep all that ink and blood to ourselves. As Plato asked, 'Are these things good for any other reason except that they end in pleasure, and get rid of and avert pain? Are you looking to any other standard but pleasure and pain when you call them good?'[19] Indeed, feelings don't just matter – they are what mattering *means*. We would expect any creature that feels pain when burned and pleasure when fed to call burning and eating *bad* and *good*, respectively, just as we would expect an asbestos creature with no digestive tract to find such designations arbitrary. Moral philosophers have tried for centuries to find some other way to define *good* and *bad*, but none has ever convinced the rest (or me). We cannot say that something is good unless we can say what it is good *for*, and if we examine all the many objects and experiences that our species calls good and ask what they are good *for*, the answer is clear: by and large, they are good for making us feel happy.

Given the importance of feelings, it would be nice to be able to say precisely what they are and how one might measure them. As we have seen, we can't do that with the kind of precision that scientists covet. Nonetheless, if the methodological and conceptual tools that science has developed do not allow us to measure the feelings of a single individual with pinpoint accuracy, they at least allow us to go stumbling in the dark with pickled rulers to measure dozens of individuals again and again. The problem facing us is a difficult one, but it is too important to ignore: why do we so often fail to know what will make us happy in the future? Science offers some intriguing answers to this question, and now that we have a sense of the problem and a general method for solving it, we are ready to inspect them.

PART III

Realism

realism (rī-ăliz´m)
The belief that things are in reality
as they appear to be in the mind.

CHAPTER 4

In the Blind Spot of the Mind's Eye

And as imagination bodies forth
The forms of things unknown, the poet's pen
Turns them to shapes, and gives to airy nothing
A local habitation and a name.

Shakespeare, *A Midsummer Night's Dream*

THIS MUCH WE KNOW for sure: Adolph Fischer did not organize the riot. He did not incite the riot. In fact, he was nowhere near the riot the night the policemen were killed. But his labour union had challenged the stranglehold that Chicago's powerful industrialists had on the men, women and children who toiled in their sweatshops toward the end of the nineteenth century, and that union needed to be taught a lesson. So Adolph Fischer was tried and, on the basis of paid and perjured testimony, sentenced to die for a crime he did not commit. On 11 November 1887, he stood on the gallows and surprised everyone with his last words: 'This is the happiest moment of my life.' A few seconds later the trapdoor beneath his feet fell away, the rope snapped his neck, and he was dead.[1]

Fortunately, his dreams of equity in the American workplace were not so easily exterminated. One year after Fischer was hanged, a bright young fellow perfected the process of dry photography, launched his revolutionary Kodak camera and instantly became one of the richest men in the world. In the decades that followed,

George Eastman developed a revolutionary management philosophy as well, giving his employees shorter hours, disability benefits, retirement annuities, life insurance, profit sharing and, ultimately, one third of the stock in his company. On 14 March 1932, the beloved inventor and humanitarian sat down at his desk, wrote a brief note, neatly capped his fountain pen and smoked a cigarette. Then he surprised everyone by killing himself.[2]

Fischer and Eastman are a fascinating contrast. Both men believed that common labourers have a right to fair wages and decent working conditions, and both dedicated much of their lives to bringing about social change at the dawn of the industrial age. Fischer failed abysmally and died a criminal, poor and reviled. Eastman succeeded absolutely and died a champion, affluent and venerated. So why did a poor man who had accomplished so little stand happily at the threshold of his own lynching while a rich man who had accomplished so much felt driven to take his own life? Fischer's and Eastman's reactions to their respective situations seem so contrary, so completely inverted, that one is tempted to chalk them up to false bravado or mental aberration. Fischer was apparently happy on the last day of a wretched existence, Eastman was apparently unhappy on the last day of a fulfilling life, and we know full well that if *we* had been standing in either of their places, *we* would have experienced precisely the opposite emotions. So what was wrong with these guys? I will ask you to consider the possibility that there was nothing wrong with them but that there *is* something wrong with you. And with me too. And the thing that's wrong with both of us is that we make a systematic set of errors when we try to imagine 'what it would feel like if'.

Imagining 'what it would feel like if' sounds like a fluffy bit of daydreaming, but in fact, it is one of the most consequential mental acts we can perform, and we perform it every day. We make decisions about whom to marry, where to work, when to reproduce, where to retire, and we base these decisions in large measure on our beliefs about how it would feel if *this* event happened but *that* one didn't.[3] Our lives may not always turn out as we wish or as we plan, but we are confident that if they had, then our happiness would have been unbounded and our sorrows thin and fleeting. Perhaps it is true that we can't always get what we want, but at least we feel sure that we know what to want in the first place. We know that happiness is to be found on the golf course and not on the assembly line, with Lana but not with Lisa, as a potter but not as a plumber, in Atlanta but not in Afghanistan, and we know these things because we can look forward in time and simulate worlds that do not yet exist. Whenever we find ourselves on the front end of a decision – *Should I have another fish stick or go directly for the Ding Dongs? Accept the job in Kansas City or stay put and hope for a promotion? Have the knee surgery or try the physical therapy first?* – we imagine the futures that our alternatives provide and then imagine how we would feel in each of them ('If the surgery didn't work, I'd always regret not having given the physical therapy a chance'). And we don't have to imagine very hard to know that we would be happier as the CEO of a FTSE 100 company than as the deadweight on a hangman's rope. Because we are the ape that looked forward, we don't actually have to *live* Adolph Fischer's or George Eastman's lives to know how it would feel to walk a mile in their shoes.

There's just one catch: the owners of the shoes didn't seem to agree with our conclusions. Fischer claimed to be happy, Eastman acted like a man who was not, so unless these guys were wrong about how it felt to live their own lives, we are forced to consider the possibility that the mistake is ours – that when we tried to imagine how it would feel to be in Fischer's or Eastman's situations, our imaginations failed us in some curious way. We are forced to consider the possibility that what clearly seems to be the better life may actually be the worse life and that when we look down the time line at the different lives *we* might lead, we may not always know which is which. We are forced to consider the possibility that we did something fundamentally wrong when we mentally slipped out of our shoes and into theirs, and that this fundamental mistake can cause us to choose the wrong future.

What might this mistake be? Imagination is a powerful tool that allows us to conjure images from 'airy nothing'. But like all tools, this one has its shortcomings, and in this and the next chapter I'll tell you about the first of them. The best way to understand this particular shortcoming of *imagination* (the faculty that allows us to see the future) is to understand the shortcomings of *memory* (the faculty that allows us to see the past) and *perception* (the faculty that allows us to see the present). As you will learn, the shortcoming that causes us to misremember the past and misperceive the present is the very same shortcoming that causes us to misimagine the future. That shortcoming is caused by a trick that your brain plays on you every minute of every hour of every day – a trick that your brain is playing on you right now. Let me tell you the brain's dirty little secret.

Little Big Head

There is a marvelous moment in most of the early Marx Brothers films in which Harpo, the cherubic mime, reaches deep into the folds of his floppy trench coat and pulls out a flügelhorn, a steaming cup of coffee, a bathroom sink or a sheep. By the age of three, most of us have learned that big things can't go inside little things, and that understanding is violated to comic effect when someone pulls plumbing or livestock from his pockets. How can a flügelhorn fit inside a raincoat? How can that tiny car hold all those merry clowns? How can the magician's assistant get folded up inside that little box? They can't, of course, and we know that, which is why we are so appreciative of the illusion that they do.

Filling in Memory

The human brain creates a similar illusion. If you've ever tried to store a full season of your favourite television show on your computer's hard drive, then you already know that faithful representations of things in the world require gobs of space. And yet, our brains take millions of snapshots, record millions of sounds, add smells, tastes, textures, a third spatial dimension, a temporal sequence, a continuous running commentary – and they do this all day, every day, year after year, storing these representations of the world in a memory bank that seems never to overflow and yet allows us to recall at a moment's notice that awful day in the sixth grade when we teased Phil Meyers about his braces and he promised to beat us up after school. How do we cram the vast universe of our experience into the relatively small storage compartment between our ears? We do what Harpo did: we cheat. As you learned in the previous chapters, the elaborate tapestry of our experience is not stored in memory – at least not in its entirety. Rather, it is compressed for storage by first being reduced to a few critical threads, such as a summary phrase ("Dinner was disappointing") or a small set of key features (tough steak, corked wine, snotty waiter). Later, when we want to remember our experience, our brains quickly reweave the tapestry by fabricating – not by actually retrieving – the bulk of the information that we experience as a memory.[4] This fabrication happens so quickly and effortlessly that we have the illusion (as a good magician's audience always does) that the entire thing was in our heads the entire time.

But it wasn't, and that fact can be easily demonstrated. For example, volunteers in one study were shown a series of slides depicting a red car as it cruises toward a give way sign, turns right and then knocks over a pedestrian.[5] After seeing the slides, some of the volunteers (the no-question group) were not asked any questions, and the remaining volunteers (the question group) were. The question these volunteers were asked was this: "Did another car pass the red car while it was stopped at the stop sign?" Next, all the volunteers were shown two pictures – one in which the red car was approaching a give way sign and one in which the red car was approaching a stop sign – and were asked to point to the picture they had actually seen. Now, if the volunteers had stored their experience in memory, then they should have pointed to the picture of the car approaching the give way sign, and indeed, more than 90 percent of the volunteers in the no-question group did just that. But 80 percent of the volunteers in the question group pointed to the picture of the car approaching a stop sign. Clearly, the question changed the volunteers' memories of their earlier experience, which is precisely what one would expect if their brains were *reweaving* their experiences – and precisely what one would *not* expect if their brains were *retrieving* their experiences.

This general finding – that information acquired *after* an event alters memory *of* the event – has been replicated so many times in so many different laboratory and field settings that it has left most scientists convinced of two things.[6] First, the act of remembering involves 'filling in' details that were not actually stored; and second, we generally cannot tell when we are doing this because filling in happens quickly and unconsciously.[7] Indeed, this phenomenon is so powerful that it happens even when we know someone is trying to trick us. For example, read the list of words below, and when you've finished, quickly cover the list with your hand. Then I will trick you.

Bed
Rest
Awake
Tired
Dream
Wake
Snooze
Blanket
Doze
Slumber
Snore
Nap
Peace
Yawn
Drowsy

Here's the trick. Which of the following words was not on the list? *Bed*, *doze*, *sleep* or *gasoline*? The right answer is *gasoline*, of course. But the other right answer is *sleep*, and if you don't believe me, then you should lift your hand from the page. (Actually, you should lift your hand from the page in any case, as we really need to move along.) If you're like most people, you knew *gasoline* was not on the list, but you mistakenly remembered reading the word *sleep*.[8] Because all the words on the list are so closely related, your brain stored the gist of what you read ('a bunch of words about sleeping') rather than storing every one of the words. Normally this would be a clever and economical strategy for remembering. The gist would serve as an instruction that enabled your brain to re-weave the tapestry of your experience and allow you to 'remember' reading the words you saw. But in this case, your brain was tricked by the fact that the gist word – the key word, the essential word – was not actually on the list. When your brain rewove the tapestry of your experience, it mistakenly included a word that was implied by the gist but that had not actually appeared, just as volunteers in the previous study mistakenly included a stop sign that was implied by the question they had been asked but that had not actually appeared in the slides they saw.

This experiment has been done dozens of times with dozens of different word lists, and these studies have revealed two surprising findings. First, people do not vaguely recall seeing the gist word and they do not simply guess that they saw the gist word. Rather, they vividly remember seeing it and they feel completely confident that it appeared.[9] Second, this phenomenon happens even when people are warned about it beforehand.[10] Knowing that a researcher is trying to trick you into falsely recalling the appearance of a gist word does not stop that false recollection from happening.

Filling in Perception

The powerful and undetectable filling in that suffuses our remembrances of things past pervades our perceptions of things present as well. For instance, if on one particularly slow Tuesday you took it upon yourself to dissect your eyeball, you would eventually come across a spot on the back of your retina where your optic nerve leaves your eye and wends its way toward your brain. The eyeball cannot register an image at the point at which the optic nerve attaches, and hence that point is known as the *blind spot*. No one can see an object that appears in the blind spot because there are no visual receptors there. And yet, if you look out into your living room, you do not notice a black hole in the otherwise smooth picture of your brother-in-law sitting on the sofa, devouring cheese dip. Why? Because your brain uses information from the areas around the blind spot to make a reasonable guess about what the blind spot would see if only it weren't blind, and then your brain fills in the scene with this information. That's right, it invents things, creates things, makes stuff up! It doesn't consult you about this, doesn't seek your approval. It just makes its best guess about the nature of the missing information and proceeds to fill in the scene – and the part of your visual experience of your cheese-dipping brother-in-law that is caused by real light reflecting off of his real face and the part that your brain just made up look *exactly alike* to you. You can convince yourself of this by closing your left eye, focusing your right eye on the magician in figure 8, and then bringing the book slowly toward you. Stay focused on the magician, but notice that when the earth moves into your blind spot, it seems to disappear. You will suddenly see whiteness where the earth actually is because your brain sees whiteness all around the earth and thus mistakenly assumes there is whiteness in your blind spot as well. If you keep moving the book toward you, the earth will reappear. Eventually, of course, your nose will touch the rabbit and you will commit an unnatural act.

Fig. 8. If you stare at the magician with your right eye and move the book slowly toward your nose, the earth will disappear into your blind spot.

The filling-in trick is not limited to the visual world. Researchers tape-recorded the sentence *The state governors met with their respective legislatures convening in the capital city.* Then they doctored the tape, substituting a cough for the first *s* in *legislatures.*[11] Volunteers heard the cough all right, but they heard it happening *between* the words because they heard the missing *s* too. Even when they were specifically instructed to listen for the missing sound, and even when they were given thousands of trials of practice, volunteers were unable to name the missing letter that their brains knew ought to be there and had thus helpfully supplied.[12] In an even more remarkable study, volunteers listened to a recording of the word *eel* preceded by a cough (which I'll denote with *). The volunteers heard the word *peel* when it was embedded in the sentence 'The *eel was on the orange' but they heard the word *heel* when it was embedded in the sentence 'The *eel was on the shoe.'[13] This is a striking finding because the two sentences differ only in their final word, which means that volunteers' brains had to wait for the last word of the sentence before they could supply the information that was missing from the second word. But they did it, and they did it so smoothly and quickly that volunteers actually *heard* the missing information being spoken in its proper position.

Experiments such as these provide us with a backstage pass that allows us to see how the brain does its marvellous magic act. Of course, if you went backstage at a magic show and got a good look at all the wires, mirrors and trapdoors, the show would be spoiled when you returned to your seat. After all, once you know how a trick works, you can't fall for it, right? Well, if you go back and try the trick in figure 8 again, you will notice that despite the detailed scientific understanding of the visual blind spot that you acquired in the last few pages, the trick still works just fine. Indeed, no matter how much you learn about optics and no matter how long you spend nosing up to that rabbit, the trick will never fail. How can that be? I have tried to convince you that things are not always as they appear. Now let me try to convince you that you can't help but believe that they are.

The Meat Loaf of Oz

Unless you skipped over childhood and went straight from strained carrots to mortgage payments, you probably remember the scene in the book *The Wonderful Wizard of Oz* in which Dorothy and her pals are cowering before the great and terrible Oz, who appears menacingly as a giant floating head. Toto the dog suddenly breaks loose, knocks over a screen in the corner of the room and reveals a little man working the controls of a machine. The protagonists are astonished, and the Scarecrow accuses the little man of being a humbug.

> 'Exactly so!' declared the little man, rubbing his hands together as if it pleased him. 'I am a humbug.' . . .
>
> 'Doesn't anyone else know you're a humbug?' asked Dorothy.
>
> 'No one knows it but you four – and myself,' replied Oz. 'I have fooled everyone so long that I thought I should never be found out.' . . .
>
> 'But, I don't understand,' said Dorothy, in bewilderment. 'How was it that you appeared to me as a great Head?'
>
> 'That was one of my tricks,' answered Oz. . . .
>
> 'I think you are a very bad man,' said Dorothy.
>
> 'Oh, no, my dear; I'm really a very good man, but I'm a very bad Wizard.'[14]

Discovering Idealism

Toward the end of the eighteenth century, philosophers had approximately the same sort of eye-opening experience that Dorothy had, and concluded (with some reluctance) that while the human brain is a very good organ, it is a very bad wizard. Prior to that time, philosophers had thought of the senses as conduits that allowed information about the properties of objects in the world to travel from the object and into the mind. The mind was like a movie screen in which the object was rebroadcast. The operation broke down on occasion, hence people occasionally saw things as they were not. But when the senses were working properly, they showed what was there. This theory of *realism* was described in 1690 by the philosopher John Locke:

> When our senses do actually convey into our understandings any idea, we cannot but be satisfied that there doth something at that time really exist without us, which doth affect our senses, and by them give notice of itself to our apprehensive faculties, and actually produce that idea which we then

perceive: and we cannot so far distrust their testimony, as to doubt that such collections of simple ideas as we have observed by our senses to be united together, do really exist together.[15]

In other words, brains believe, but they don't *make* believe. When people see giant floating heads, it is because giant heads are actually floating in their purview, and the only question for a psychologically minded philosopher was how brains accomplish this amazing act of faithful reflection. But in 1781 a reclusive German professor named Immanuel Kant broke loose, knocked over the screen in the corner of the room and exposed the brain as a humbug of the highest order. Kant's new theory of *idealism* claimed that our perceptions are not the result of a physiological process by which our eyes somehow transmit an image of the world into our brains, but rather, they are the result of a psychological process that combines what our eyes see with what we already think, feel, know, want and believe, and then uses this combination of sensory information and preexisting knowledge to construct our perception of reality. 'The understanding can intuit nothing, the senses can think nothing,' Kant wrote. 'Only through their union can knowledge arise.'[16] The historian Will Durant performed the remarkable feat of summarizing Kant's point in a single sentence: 'The world as we know it is a construction, a finished product, almost – one might say – a manufactured article, to which the mind contributes as much by its moulding forms as the thing contributes by its stimuli.'[17] Kant argued that a person's perception of a floating head is *constructed* from the person's knowledge of floating heads, memory of floating heads, belief in floating heads, need for floating heads and sometimes – but not always – from the actual presence of a floating head itself. Perceptions are portraits, not photographs, and their form reveals the artist's hand every bit as much as it reflects the things portrayed.

This theory was a revelation, and in the centuries that followed, psychologists extended it by suggesting that each individual makes roughly the same journey of discovery that philosophy did. In the 1920s, the psychologist Jean Piaget noticed that the young child often fails to distinguish between her perception of an object and the object's actual properties, hence she tends to believe that things really are as they appear to be – and that others must therefore see them as she does. When a two-year-old child sees her playmate leave the room, and then sees an adult remove a cookie from a cookie jar and hide it in a drawer, she expects that her playmate will later look for the cookie in the drawer – despite the fact that her playmate was not in the room when the adult moved the cookie to the drawer from the jar.[18] Why? Because the two-year-old child knows the cookie is in the drawer and thus expects that everyone else knows this as well. Without a distinction between *things in the world* and *things in the mind*, the child cannot understand how different minds can contain different things. Of course, with increasing maturity, children shift from realism to idealism, coming to realize that perceptions are merely points of view, that what they see is not necessarily what there is, and that two people may thus have different perceptions of or beliefs about the same thing. Piaget concluded that 'the child is a realist in its thought' and that 'its progress consists in ridding itself of this initial realism'.[19] In other words, like philosophers, ordinary people start out as realists but get over it soon enough.

Escaping Realism

But if realism goes away, it doesn't get very far. Research shows that even adults act like realists under certain circumstances. For example, in one study, a pair of adult volunteers were seated on opposite sides of a set of cubbyholes, as shown in figure 9.[20] Some common objects were placed in several of the cubbies. Some of these cubbies were open on both sides, so that items such as the large truck and the medium truck were clearly visible to both volunteers. Other cubbies were open on only one side, so that items such as the small truck could be seen by one volunteer but not by the other. The volunteers played a game in which the person with the occluded view (the director) told the person with the clear view (the mover) to move certain objects to certain locations. Now, what should have happened when the director said, 'Move the small truck to the bottom row'? If the mover were an idealist, she would move the medium truck because she would realize that the director could not see the small truck, hence he must have been referring to the medium truck, which, from *the director's* point of view, was the smallest. On the other hand, if the mover were a realist, then she would move the small truck without regard for the fact that the director could not see it as she could, hence could not have been referring to it when he gave his instruction. So which truck did the movers actually move?

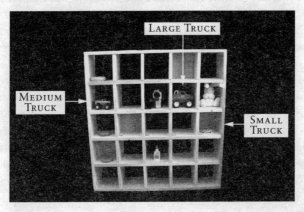

Fig. 9.

The medium truck, of course. What – did you think they were stupid? These were normal adults. They had intact brains, good jobs, bank accounts, nice table manners – all the usual stuff. They knew that the director had a different point of view and thus must have been talking about the medium truck when he said, 'Move the small truck.' But while these normal adults with intact brains *behaved* like perfect idealists, their hands told only half the story. In addition to measuring the mover's hand movements, the researchers used an eye tracker to measure the mover's eye movements as well. The eye tracker revealed that the moment the mover heard the phrase 'Move the small truck,' she briefly looked toward the small truck – *not* the medium truck, which was the smallest truck the director could see, but the small truck, which was the smallest truck that *she* could see. In other words, the mover's brain initially interpreted the phrase 'the small truck' as a reference to the smallest truck *from her own point of view*, without regard for the fact that the director's point of view was different. Only after briefly flirting with the idea of moving the small truck did the mover's brain consider the fact that the director had a different view and thus must have meant the medium truck, at which point her brain sent her hand instructions to move the proper truck. The hand behaved like an idealist, but the eye revealed that the brain was a momentary realist.

Experiments such as these suggest that we do not outgrow realism so much as we learn to outfox it, and that even as adults our perceptions are characterized by an initial moment of realism.[21] According to this line of reasoning, we automatically assume that our subjective experience of a thing is a faithful representation of the thing's properties. Only later – if we have the time, energy and ability – do we rapidly repudiate that assumption and consider the possibility that the real world may not actually be as it appears to us.[22] Piaget described realism as 'a spontaneous and immediate tendency to confuse the sign and the thing signified',[23] and research shows that this tendency to equate our subjective sense of things with the objective properties of those things remains spontaneous and immediate throughout our lives. It does not go away forever, and it does not go away on occasion. Rather, it is brief, unarticulated and rapidly unravelled, but it is always the first step in our perception of the world. We believe what we see, and then unbelieve it when we have to.

All of this suggests that the psychologist George Miller was right when he wrote, 'The crowning intellectual accomplishment of the brain is the real world.'[24] The three-and-a-half-pound meat loaf between our ears is not a simple recording device but a remarkably smart computer that gathers information, makes shrewd judgments and even shrewder guesses, and offers us its best interpretations of the way things are. Because those interpretations are usually so good, because they usually bear such a striking resemblance to the world as it is actually constituted, *we do not realize that we are seeing an interpretation*. Instead, we feel as though we are sitting comfortably inside our heads, looking out through the clear glass windshield of our eyes, watching the world as it truly is. We tend to forget that our brains are talented forgers, weaving a tapestry of memory and perception whose detail is so compelling that its inauthenticity is rarely detected. In a sense, each of us is a counterfeiter who prints phony dollar bills and then happily accepts them for payment, unaware that he is both the perpetrator and victim of a well-orchestrated fraud. As you are about to see, we sometimes pay a steep price for allowing ourselves to lose sight of this fundamental fact, because the mistake we make when we momentarily ignore the filling-in trick and unthinkingly accept the validity of our memories and our perceptions is precisely the same mistake we make when we imagine our futures.

An Embarrassment of Tomorrows

When John Lennon asked us to 'imagine there's no countries', he was quick to add, 'it isn't hard to do'. Indeed, imagination is usually effortless. When we think about the pastrami on rye that we intend to have for lunch, or the new pair of flannel pajamas that Mum swears she mailed last week, we do not have to set aside a block of time between other appointments, roll up our sleeves and get down to the serious work of conjuring up images of sandwiches and sleepwear. Rather, the moment we have the slightest inclination to consider these things, our brains effortlessly use what they know about delis and lunches and parcels and mothers to construct mental pictures (warm pastrami, dark rye, tartan-plaid pajamas with bunny feet) that we experience as the products of imagination. Like perceptions and memories, these mental pictures pop into our consciousness *fait accompli*. We should be grateful for the ease with which our imaginations provide this useful service, but because we do not consciously supervise the construction of these mental images, we tend to treat them as we treat memories and perceptions – initially assuming that they are *accurate representations* of the objects we are imagining.

For instance, right now you can probably imagine a plate of spaghetti and tell me how much you would enjoy eating it for dinner tomorrow evening. Fine. Now notice two things. First, this wasn't particularly taxing. You could probably imagine pasta all day long without ever breaking a sweat, letting your brain do the heavy construction work while you lounge around in your new pyjamas. Second, notice that the spaghetti you imagined was much richer than the spaghetti I asked you about. Perhaps your imaginary spaghetti was the goopy slop from the can, or perhaps it was fresh basil-rosemary pasta topped with a silky bolognese. The sauce could have been tomato, cream, clam or even grape jelly. The noodles might have been piled beneath a pair of traditional meatballs, or sprinkled with a half-dozen slivers of duck sausage studded with capers and pine nuts. Maybe you imagined eating the spaghetti while standing at your kitchen counter with a newspaper in one hand and a Coke in the other, or maybe you imagined that your waiter had given you the small table near the fireplace at your favourite trattoria and poured you a fat 1990 Barolo to start. Whatever you imagined, it's a pretty good bet that when I said *spaghetti*, you did not have an unrequited urge to interrogate me about the nuances of sauce and locale before envisioning a single noodle. Instead, your brain behaved like a portrait artist commissioned to produce a full-colour oil from a rough charcoal sketch, filling in all the details that were absent from my question and serving you a particularly heaped helping of imaginary pasta. And when you estimated your enjoyment of this future spaghetti, you responded to this particular mental image as you respond to particular memories and particular perceptions – as though the details had been specified by the thing you were imagining rather than fabricated by your brain.

In so doing, you made an error that your future spaghetti-eating self may regret.[25] The phrase 'spaghetti for dinner tomorrow evening' does not describe an event so much as it describes a family of events, and the particular member of the family that you imagined influenced your predictions about how much you'd enjoy eating it. Indeed, trying to predict how much you will enjoy a plate of spaghetti without knowing *which* plate of spaghetti is like trying to predict how much you will pay for a car without knowing *which* car (Ferrari or Chevy?), or trying to predict how proud you will be of your spouse's accomplishment without knowing *which* accomplishment (winning a Nobel Prize or finding the best divorce lawyer in the city?), or trying to predict how sad you will be when a relative dies without knowing *which* relative (dear old Dad or cranky Great Uncle Sherman on Cousin Ida's side twice removed?). There are endless variations on spaghetti, and the particular variation you imagined surely influenced how much you expected to enjoy the experience. Because these details are so crucial to an accurate prediction of your response to the event you were imagining, and because these important details were not known, you would have been wise to withhold your prediction about spaghetti, or at least to temper it with a disclaimer such as 'I expect to like the spaghetti *if* it is al dente with smoked pomodoro.'

But I'm willing to bet that you didn't withhold, you didn't disclaim, and that you instead conjured up a plate of imaginary spaghetti faster than Delia Smith on rollerskates, then made a confident prediction about the relationship you expected to have with that food. If you didn't do that, then congratulations. Give yourself a medal. But if you did, then know you are not alone. Research suggests that when people make predictions about their reactions to future events, they tend to neglect the fact that their brains have performed the filling-in trick as an integral part of the act of imagination.[26] For example, volunteers in one study were asked to predict what they would do in a variety of future situations – how much time they would be willing to spend answering questions in a telephone survey, how much money they would be willing to spend to celebrate a special occasion at a restaurant in San Francisco, and so on.[27] Volunteers also reported how confident they were that each of these predictions was correct. Before making predictions, some volunteers were asked to describe all the details of the future event they were imagining ('I am imagining eating wine-braised short ribs with roasted root vegetables and parsley coulis at Jardiniere') and were then told that they should assume that each of these details was perfectly accurate (the assumers). Other volunteers were not asked to describe these details and were not told to make any assumptions (nonassumers). The results showed that nonassumers were every bit as confident as assumers were. Why? Because when they were asked about *dinner*, the nonassumers quickly and unconsciously generated a mental image of a particular dish at a particular restaurant, and then assumed that these details were accurate, rather than having been conjured from airy nothing.

We all find ourselves in the same fix from time to time. Our spouse asks us to attend a party next Friday night, our brains instantly manufacture an image of a cocktail party in the penthouse of a downtown hotel with waiters in black tie carrying silver trays of hors d'oeuvres past a slightly bored harpist, and we predict our reaction to the imagined event with a yawn that sets new records for duration and jaw extension. What we generally fail to consider is how many different kinds of parties there are – birthday celebrations, gallery openings, cast parties, yacht parties, office parties, orgies, wakes – and how different our reactions would be to each. So we tell our spouse that we'd rather skip the party, our spouse naturally drags us along anyhow and we have a truly marvellous time. Why? Because the party involved cheap beer and hula hoops rather than classical music and seaweed crackers. It was precisely our style, and we liked what we predicted we'd hate because our prediction was based on a detailed image that reflected our brain's best guess, which was in this case dead wrong. The point here is that when we imagine the future, we often do so in the blind spot of our mind's eye, and this tendency can cause us to misimagine the future events whose emotional consequences we are attempting to weigh.

This tendency transcends mundane predictions about parties, restaurants and plates of spaghetti. For example, most of us have no doubt that we would enjoy being an Eastman more than we would enjoy being a Fischer – no doubt, that is, unless we pause to consider how quickly and unquestioningly our brains filled in the details of their lives and deaths, and how much all those made-up details mattered. Consider a pair of stories that your brain almost certainly did not invent for you at the beginning of this chapter.

You are a young German immigrant who lives in the teeming, dirty city that is nineteenth-century Chicago. A few wealthy families – the Armours, McCormicks, Swifts and Fields – have monopolized their industries and have the right to use you and your family as they would use machines and horses. You devote your time to a small newspaper whose editorials call for social justice, but you are no fool, and you know that these essays will change nothing and that the factories will churn on, producing paper, producing pork, producing tractors, and spitting out the tired workers whose blood and sweat feed the engines of production. You are dispensable and insignificant. Welcome to America. One evening an altercation breaks out between some factory workers and the local police in Haymarket Square, and although you were not present when the bomb was thrown, you are rounded up with other 'anarchist leaders' and charged with masterminding a riot. Suddenly your name is on the front page of every major newspaper and you have a national platform for your opinions. When the judge sentences you on the basis of fabricated evidence, you realize that this ignominious moment will be preserved in history books, that you will be known as 'the Haymarket Martyr', and that your execution will pave the way for the reforms you sought but were impotent to establish. A few decades from now, there will be a far better America than this one, and its citizens will honour you for your sacrifice. You are not a religious man, but you cannot help but think for a moment of Jesus on the cross – falsely accused, unjustly convicted and cruelly executed – giving his life so that a great idea might live in the centuries to come. As you prepare to die you feel nervous, of course. But in some deep sense, this moment is a stroke of luck, the culmination of a dream – perhaps, you might even say, the happiest moment of your life.

Cut to a second story: Rochester, New York, 1932, the midst of the Great Depression. You are a seventy-seven-year-old man who has spent his life building empires, advancing technology and using his wealth to endow libraries, symphonies, colleges and dental clinics that have improved the lives of millions. The happiest moments of your long life were spent tinkering with a camera, touring Europe's art museums, fishing, hunting or doing carpentry in your lodge in North Carolina. But spinal disease has made it increasingly difficult for you to lead the active life you've always enjoyed, and every day you spend in bed is a sad mockery of the vibrant man you once were. You will never get younger, you will never get better. The good days are over, and more days merely mean more decrepitude. One Monday afternoon you sit down at your desk, uncap your favourite fountain pen, and write these words on a legal pad: 'Dear friends: My work is done. Why wait?' Then you light a cigarette and, when you've enjoyed the last of it, stub it out and carefully place the nose of your Luger automatic against your chest. Your physician showed you how to locate your heart, and now you can feel it beating rapidly beneath your hand. As you prepare to pull the trigger you feel nervous, of course. But in some deep sense you know that this one well-aimed bullet will allow you to leave a beautiful past and escape a bitter future.

Okay, lights up. These details about Adolph Fischer's and George Eastman's lives are accurate, but that's not really the point. The point is that just as there are parties and pastas you like and parties and pastas you don't, there are ways of being rich and ways of being executed that make the former less marvellous and the latter less awful than we might otherwise expect. One reason why you found Fischer's and Eastman's reactions so perverse is that you almost certainly misimagined the details of their situations. And yet, without a second thought, you behaved like an unrepentant realist and confidently based your predictions about how you would feel on details that your brain had invented while you weren't watching. Your mistake was not in imagining things you could not know – that is, after all, what imagination is for. Rather, your mistake was in unthinkingly treating what you imagined as though it were an accurate representation of the facts. You are a very fine person, I'm sure. But you are a very bad wizard.

Onward

If you'd been given your choice of brains at the moment of conception, you probably wouldn't have chosen the tricky one. Good thing no one asked you. Without the filling-in trick you would have sketchy memories, an empty imagination and a small black hole following you wherever you went. When Kant wrote that 'perception without conception is blind',[28] he was suggesting that without the filling-in trick we would have nothing even remotely resembling the subjective experience that all of us take for granted. We see things that aren't really there and we remember things that didn't really happen, and while these may sound like symptoms of mercury poisoning, they are actually critical ingredients in the recipe for a seamlessly smooth and blessedly normal reality. But that smoothness and normality come at a price. Even though we are aware in some vaguely academic sense that our brains are doing the filling-in trick, we can't help but expect the future to unfold with the details we imagine. As we are about to see, the details that the brain puts in are not nearly as troubling as the details it leaves out.

CHAPTER 5

The Hound of Silence

O hateful Error, Melancholy's child,
Why dost thou show to the apt thoughts of men
The things that are not?

Shakespeare, *Julius Caesar*

SILVER BLAZE HADN'T BEEN MISSING for long when Inspector Gregory and Colonel Ross identified the stranger who had sneaked into the stable and stolen the prize racehorse. But as usual, Sherlock Holmes was one step ahead of the police. The colonel turned to the great detective:

'Is there any point to which you would wish to draw my attention?'
'To the curious incident of the dog in the night-time.'
'The dog did nothing in the night-time.'
'That was the curious incident,' remarked Sherlock Holmes.[1]

It seems that a dog lived in the stable, that both of the stable hands had slept through the theft, and that these two facts had allowed Holmes to make one of his indubitably shrewd deductions. As he later explained:

I had grasped the significance of the silence of the dog. . . . A dog was kept in the stables, and yet, though some one had been in and had fetched out a horse, he had not barked enough to arouse the two lads in the loft. Obviously the midnight visitor was some one whom the dog knew well.[2]

Although the inspector and the colonel were aware of what had happened, only Holmes was aware of what *hadn't* happened: the dog hadn't barked, which meant that the thief was not the stranger whom the police had identified. By paying careful attention to the absence of an event, Sherlock Holmes further distinguished himself from the rest of humankind. As we are about to see, when the rest of humankind imagines the future, it rarely notices what imagination has missed – and the missing pieces are much more important than we realize.

The Sailors Not

If you live in a city with tall buildings, then you already know that pigeons have an uncanny ability to defecate at precisely the moment, speed and position required to score a direct hit on your most expensive sweater. Given their talent as bombardiers, it seems odd that pigeons can't learn to do much simpler things. For example, if a pigeon is put in a cage with two levers that can be briefly illuminated, it can easily learn to press the illuminated lever to get a reward of bird seed – but it can *never* learn to press the unilluminated lever to receive the same reward.[3] Pigeons have no trouble figuring out that the presence of a light signals an opportunity for eating, but they cannot learn the same thing about the *absence* of a light. Research suggests that human beings are a bit like pigeons in this regard. For example, volunteers in one study played a deduction game in which they were shown a set of trigrams (i.e., three-letter combinations such as SXY, GTR, BCG and EVX). The experimenter then pointed to one of the trigrams and told the volunteers that *this* trigram was special. The volunteers' job was to figure out what made the trigram special – that is, to figure out which feature of the special trigram distinguished it from the others. Volunteers saw set after set, and each time the experimenter pointed out the special one. How many sets did volunteers have to see before they deduced the distinctive feature of the special trigram? For half the volunteers, the special trigram was distinguished by the fact that it and only it contained the letter *T*, and these volunteers needed to see about thirty-four sets of trigrams before they figured out that the presence of *T* is what made a trigram special. For the other half of the volunteers, the special trigram was always distinguished by the fact that it and only it *lacked* the letter *T*. The results were astounding. No matter how many sets of trigrams they saw, none of the volunteers *ever* figured this out.[4] It was easy to notice the presence of a letter but, like the barking of a dog, it was impossible to notice its absence.

Absence in the Present

If this tendency were restricted to bird seed and trigrams, we wouldn't care much about it. But as it turns out, the general inability to think about absences is a potent source of error in everyday life. For example, a moment ago I suggested that pigeons have an unusual talent for hitting pedestrians, and if you've ever been the victim of a really good splotching, you've probably concluded the same thing. But what makes us think that the pigeons are actually taking aim and hitting what they aim for? The answer is that most of us can remember far too many instances in which we momentarily passed beneath a ledge festooned with those obnoxious flying rats and were squarely nailed by a stinky white dollop, despite the fact that from the air a human head constitutes a relatively small and fast-moving target. Fair enough. But if we really want to know whether the pigeons are out to get us and have the requisite skills to do so, we must also consider the times when we walked beneath that ledge and came away clean. The *right* way to calculate the animosity and marksmanship of the urban pigeon is to consider *both* the presence and the absence of poo on our jackets. If the pigeons have hit us nine times out of ten, then we should probably give them credit for their accuracy as well as a wide berth, but if they've hit us nine times out of nine thousand, then what seems like good aim and bad attitude is probably nothing more than dumb luck. The misses are *crucial* to determining what kinds of inferences we can legitimately draw from the hits. Indeed, when scientists want to establish the causal relationship between two things – cloud seeding and rain, heart attacks and cholesterol, you name it – they compute a mathematical index that takes into account *co-occurrences* (how many people who *do* have high cholesterol *do* have heart attacks?) and *non-co-occurrences* (how many people who *do* have high cholesterol *do not* have heart attacks, and how many people who *do not* have high cholesterol *do* have heart attacks?) and *co-absences* (how many people who *don't* have high cholesterol *don't* have heart attacks?). All of these quantities are necessary to assess accurately the likelihood that the two things have a real causal relationship.

This is all very sensible, of course. To statisticians. But studies show that when ordinary people want to know whether two things are causally related, they routinely search for, attend to, consider and remember information about what *did* happen and fail to search for, attend to, consider and remember information about what *did not*.[5] Apparently, people have been making this mistake for a very long time. Nearly four centuries ago, the philosopher and scientist Sir Francis Bacon wrote about the ways in which the mind errs, and he considered the failure to consider absences among the most serious:

> By far the greatest impediment and aberration of the human understanding arises from [the fact that] . . . those things which strike the sense outweigh things which, although they may be more important, do not strike it directly. Hence, contemplation usually ceases with seeing, so much so that little or no attention is paid to things invisible.[6]

Bacon illustrated his point with a story (which, it turns out, he borrowed from Cicero, who told it seventeen centuries earlier) about a visitor to a Roman temple. To impress the visitor with the power of the gods, the Roman showed him a portrait of several pious sailors whose faith had presumably allowed them to survive a recent shipwreck. When pressed to accept this as evidence of a miracle, the visitor astutely inquired, 'But where are the pictures of those who perished after taking their vows?'[7] Scientific research suggests that ordinary people like us rarely ask to see pictures of the missing sailors.[8]

Our inability to think about absences can lead us to make some fairly bizarre judgments. For example, in a study done about three decades ago, Americans were asked which countries were most similar to each other – Ceylon and Nepal or West Germany and East Germany. Most picked the latter pair.[9] But when they were asked which countries were most *dissimilar*, most Americans picked the latter pair as well. Now, how can one pair of countries be both more similar *and* more dissimilar than another pair? They can't, of course. But when people are asked to judge the similarity of two countries, they tend to look for the presence of similarities (of which East and West Germany had many – for example, their names) and *ignore the absence* of similarities. When they are asked to judge the dissimilarities of two countries, they tend to look for the presence of dissimilarities (of which East and West Germany had many – for example, their governments) and *ignore the absence* of dissimilarities.

The tendency to ignore absences can befuddle more personal decisions as well. For example, imagine that you are preparing to go on a holiday to one of two islands: Moderacia (which has average weather, average beaches, average hotels and average nightlife) or Extremia (which has beautiful weather and fantastic beaches but crummy hotels and no nightlife). The time has come to make your reservations, so which one would you choose? Most people pick Extremia.[10] But now imagine that you are already holding tentative reservations for both destinations and the time has come to cancel one of them before they charge your credit card. Which would you cancel? Most people choose to cancel their reservation on Extremia. Why would people both select *and* reject Extremia? Because when we are selecting, we consider the positive attributes of our alternatives, and when we are rejecting, we consider the negative attributes. Extremia has the most positive attributes *and* the most negative attributes, hence people tend to select it when they are looking for something to select *and* they reject it when they are looking for something to reject. Of course, the logical way to select a holiday is to consider both the presence *and* the absence of positive *and* negative attributes, but that's not what most of us do.

Absence in the Future

Our inattention to absences influences the way that we think about the future. Just as we do not remember every detail of a past event (what colour socks did you wear to your high school graduation?) or see every detail of a current event (what colour socks is the person behind you wearing at this very moment?), so do we fail to imagine every detail of a future event. You could close your eyes right now and spend two full hours imagining yourself driving around in a silver Mercedes-Benz SL600 Roadster with a twin-turbocharged 36-valve 5.5-litre V-12 engine. You could imagine the curve of the front grille, the slant of the windshield and the newish smell of the black leather upholstery. But no matter how long you spent doing this, if I were then to ask you to inspect the mental image you'd created and read to me the numbers on the licence plate, you would be forced to admit that you'd left out that particular detail. No one can imagine *everything*, of course, and it would be absurd to suggest that they should. But just as we tend to treat the details of future events that we *do* imagine as though they were actually going to happen, we have an equally troubling tendency to treat the details of future events that we *don't* imagine as though they were *not* going to happen. In other words, we fail to consider how much imagination fills in, but we also fail to consider how much it leaves out.

To illustrate this point, I often ask people to tell me how they think they would feel two years after the sudden death of their eldest child. As you can probably guess, this makes me quite popular at parties. I know, I know – this is a gruesome exercise and I'm not asking you to do it. But the fact is that *if* you did it, you would probably give me the answer that almost everyone gives me, which is some variation on *Are you out of your damned mind? I'd be devastated – totally devastated. I wouldn't be able to get out of bed in the morning. I might even kill myself. So who invited you to this party anyway?* If at this point I'm not actually wearing the person's cocktail, I usually probe a bit further and ask how he came to his conclusion.

What thoughts or images came to mind, what information did he consider? People typically tell me that they imagined hearing the news, or they imagined attending the funeral, or they imagined opening the door to an empty bedroom. But in my long history of asking this question and thereby excluding myself from every social circle to which I formerly belonged, I have yet to hear a single person tell me that in addition to these heartbreaking, morbid images, they also imagined *the other things* that would inevitably happen in the two years following the death of their child. Indeed, not one person has ever mentioned attending another child's school play, or making love with his spouse, or eating a toffee apple on a warm summer evening, or reading a book, or writing a book, or riding a bicycle, or any of the many other activities that we – and that they – would expect to happen in those two years. Now, I am in no way, shape or form suggesting that a bite of gooey candy compensates for the loss of a child. That isn't the point. What I am suggesting is that the two-year period following a tragic event has to contain *something* – that is, it must be filled with episodes and occurrences of *some* kind – and these episodes and occurrences must have *some* emotional consequences. Regardless of whether those consequences are large or small, negative or positive, one cannot answer my question accurately without considering them. And yet, not one person I know has ever imagined anything other than the single, awful event suggested by my question. When they imagine the future, there is a whole lot missing, and the things that are missing matter.

This fact was illustrated by a study in which college students at the University of Virginia were asked to predict how they would feel a few days after their school's football team won or lost an upcoming game against the University of North Carolina.[11] Before making these predictions, one group of students (the describers) was asked to describe the events of a typical day, and one group of students (the nondescribers) was not. A few days later the students were asked to report how happy they actually were, and the results showed that only the nondescribers had drastically overestimated the impact that the win or the loss would have on them. Why? Because when nondescribers imagined the future, they tended to leave out details about the things that would happen after the game was over. For example, they failed to consider the fact that right after their team lost (which would be sad) they would go get drunk with their friends (which would be lovely), or that right after their team won (which would be lovely) they would have to go to the library and start studying for their chemistry final (which would be sad). The nondescribers were focused on one and only one aspect of the future – the outcome of the football game – and they failed to imagine other aspects of the future that would influence their happiness, such as drunken parties and chemistry exams. The describers, on the other hand, were more accurate in their predictions precisely because they were *forced* to consider the details that the nondescribers left out.[12]

It is difficult to escape the focus of our own attention – difficult to consider what it is we may not be considering – and this is one of the reasons why we so often mispredict our emotional responses to future events. For example, most Americans can be classified as one of two types: those who live in California and are happy they do, and those who don't live in California but believe they'd be happy if they did. Yet, research shows that Californians are actually no happier than anyone else – so why does everyone (including Californians) seem to believe they are?[13] California has some of the most beautiful scenery and some of the best weather in the continental United States, and when non-Californians hear that magic word their imaginations instantly produce mental images of sunny beaches and giant redwood trees. But while Los Angeles has a better climate than Columbus, climate is just one of many things that determine a person's happiness – and yet all those other things are missing from the mental image. If we were to add some of these missing details to our mental image of beaches and palm trees – say, traffic, supermarkets, airports, sports teams, cable rates, housing costs, earthquakes, landslides, and so on – then we might recognize that L.A. beats Columbus in some ways (better weather) and Columbus beats L.A. in others (less traffic). We think that Californians are happier than Ohioans because we imagine California with so few details – and we make no allowance for the fact that the details we are failing to imagine could drastically alter the conclusions we draw.[14]

The tendency that causes us to overestimate the happiness of Californians also causes us to underestimate the happiness of people with chronic illnesses or disabilities.[15] For example, when sighted people imagine being blind, they seem to forget that blindness is not a full-time job. Blind people can't see, but they do most of the things that sighted people do – they go on picnics, pay their taxes, listen to music, get stuck in traffic – and thus they are just as happy as sighted people are. They can't do *everything* sighted people can do, sighted people can't do *everything* that they can do, and thus blind and sighted lives are not identical. But whatever a blind person's life is like, it is about much more than blindness. And yet, when sighted people imagine being blind, they fail to imagine all the other things that such a life might be about, hence they mispredict how satisfying such a life can be.

On the Event Horizon

About fifty years ago a Pygmy named Kenge took his first trip out of the dense, tropical forests of Africa and onto the open plains in the company of an anthropologist. Buffalo appeared in the distance – small black specks against a bleached sky – and the Pygmy surveyed them curiously. Finally, he turned to the anthropologist and asked what kind of insects they were. 'When I told Kenge that the insects were buffalo, he roared with laughter and told me not to tell such stupid lies.'[16] The anthropologist wasn't stupid and he hadn't lied. Rather, because Kenge had lived his entire life in a dense jungle that offered no views of the horizon, he had failed to learn what most of us take for granted, namely, that things look different when they are far away. You and I don't mix up our insects and our ungulates because we are used to looking out across vast expanses, and we learned early on that objects make smaller images on our retinas when they are distant than when they are nearby. How do our brains know whether a small retinal image is being made by a small object that is nearby or a large object that is distant? Details, details, details! Our brains know that the surfaces of nearby objects afford fine-grained details that blur and blend as the object recedes into the distance, and thus they use the level of detail that we can see to estimate the distance between our eye and the object. If the small retinal image is detailed – we can see the fine hairs on a mosquito's head and the cellophane texture of its wings – our brains assume that the object is about an inch from our eye. If the small retinal image is not detailed – we can see only the vague contour and shadowless form of the buffalo's body – our brains assume that the object is a few thousand yards away.

Just as objects that are near to us in space appear to be more detailed than those that are far away, so do events that are near to us in time.[17] Whereas the near future is finely detailed, the far future is blurry and smooth. For example, when young couples are asked to say what they think of when they envision 'getting married', those couples who are a month away from the event (either because they are getting married a month later or because they got married a month earlier) envision marriage in a fairly abstract and blurry way, and they offer high-level descriptions such as 'making a serious commitment' or 'making a mistake'. But couples who are getting married the next day envision marriage's concrete details, offering descriptions such as 'having pictures made' or 'wearing a special outfit'.[18] Similarly, when volunteers are asked to imagine themselves locking a door the next day, they describe their mental images with detailed phrases such as 'putting a key in the lock', but when volunteers are asked to imagine themselves locking a door next year, they describe their mental images with vague phrases such as 'securing the house'.[19] When we think of events in the distant past or distant future we tend to think abstractly about *why* they happened or will happen, but when we think of events in the near past or near future we tend to think concretely about *how* they happened or will happen.[20]

Seeing in time is like seeing in space. But there is one important difference between spatial and temporal horizons. When we perceive a distant buffalo, our brains are aware of the fact that the buffalo looks smooth, vague and lacking in detail *because* it is far away, and they do not mistakenly conclude that the buffalo itself is smooth and vague But when we remember or imagine a temporally distant event, our brains seem to overlook the fact that details vanish with temporal distance, and they conclude instead that the distant events actually *are* as smooth and vague as we are imagining and remembering them. For example, have you ever wondered why you often make commitments that you deeply regret when the moment to fulfil them arrives? We all do this, of course. We agree to babysit the nephews and nieces next month, and we look forward to that obligation even as we jot it in our diary. Then, when it actually comes time to buy the Happy Meals, set up the Barbie playset, hide the bong and ignore the fact that the NBA playoffs are on at one o'clock, we wonder what we were thinking when we said yes. Well, here's what we were thinking: when we said yes we were thinking about babysitting in terms of *why* instead of *how*, in terms of causes and consequences instead of execution, and we failed to consider the fact that the detail-free babysitting we were imagining would not be the detail-laden babysitting we would ultimately experience. Babysitting next month is 'an act of love', whereas babysitting right now is 'an act of lunch', and expressing affection is spiritually rewarding in a way that buying French fries simply isn't.[21]

Perhaps it isn't surprising that the gritty details of babysitting that are so salient to us as we execute them were not part of our mental image of babysitting when we imagined it a month earlier, but what *is* surprising is how surprised we are when those details finally come into view. Distant babysitting has the same illusory smoothness that a distant cornfield does,[22] but while we all know that a cornfield isn't *really* smooth and that it just *looks* that way from a far remove, we seem only dimly aware of the same fact when it comes to events that are far away in time. When volunteers are asked to 'imagine a good day', they imagine a greater *variety* of events if the good day is tomorrow than if the good day is a year later.[23] Because a good day tomorrow is imagined in considerable detail, it turns out to be a lumpy mixture of mostly good stuff ('I'll sleep late, read the paper, go to the movies and see my best friend') with a few unpleasant chunks ('But I guess I'll also have to rake the stupid leaves'). On the other hand, a good day next year is imagined as a smooth purée of happy episodes. What's more, when people are asked how *realistic* they think these mental images of the near and far future are, they claim that the smooth purée of next year is every bit as realistic as the lumpy stew of tomorrow. In some sense, we are like pilots who land our planes and are genuinely shocked to discover that the cornfields that looked like smooth, yellow rectangles from the air are actually filled with – of all things – corn! Perception, imagination and memory are remarkable abilities that have a good deal in common, but in at least one way, perception is the wisest of the triplets. We rarely mistake a distant buffalo for a nearby insect, but when the horizon is temporal rather than spatial, we tend to make the same mistake that Pygmies do.

The fact that we imagine the near and far futures with such different textures causes us to value them differently as well.[24] Most of us would pay more to see a Broadway show tonight or to eat an apple pie this afternoon than we would if the same ticket and the same pie were to be delivered to us next month. There is nothing irrational about this. Delays are painful, and it makes sense to demand a discount if one must endure them. But studies show that when people imagine the pain of waiting, they imagine that it will be worse if it happens in the near future than in the far future, and this leads to some rather odd behaviour.[25] For example, most people would rather receive $20 in a year than $19 in 364 days because a one-day delay that takes place in the *far future* looks (from here) to be a minor inconvenience. On the other hand, most people would rather receive $19 today than $20 tomorrow because a one-day delay that takes place in the *near future* looks (from here) to be an unbearable torment.[26] Whatever amount of pain a one-day wait entails, that pain is surely the same whenever it is experienced; and yet, people imagine a near-future pain as so severe that they will gladly pay a dollar to avoid it, but a far-future pain as so mild that they will gladly accept a dollar to endure it.

Why does this happen? The vivid detail of the near future makes it much more palpable than the far future, thus we feel more anxious and excited when we imagine events that will take place soon than when we imagine events that will take place later. Indeed, studies show that the parts of the brain that are primarily responsible for generating feelings of pleasurable excitement become active when people imagine receiving a reward such as money in the near future but not when they imagine receiving the same reward in the far future.[27] If you've ever bought too many boxes of Thin Mints from the Girl Scout who hawks her wares in front of the local library but too few boxes from the Girl Scout who rings your doorbell and takes your order for future delivery, then you've experienced this anomaly yourself. When we spy the future through our prospecti-scopes, the clarity of the next hour and the fuzziness of the next year can lead us to make a variety of mistakes.

Onward

Before heading back to Baker Street, Sherlock Holmes couldn't resist polishing his own calabash and giving Inspector Gregory one last poke in the eye. Holmes confided in Watson: "See the value of imag-ination," said Holmes. "It is the one quality which Gregory lacks. We imagined what might have happened, acted upon the suppasi-tion, and find ourselves justified."[28]

A very fine poke, but not a very fair one. Inspector Gregory's problem wasn't that he lacked imagination but that he trusted it. Any brain that does the filling-in trick is bound to do the leaving-out trick as well, and thus the futures we imagine contain some details that our brains invented and lack some details that our brains ignored. The problem isn't that our brains fill in and leave out. God help us if they didn't. No, the problem is that they do this so *well* that we aren't aware it is happening. As such, we tend to accept the brain's products uncritically and expect the future to unfold with the details – and with *only* the details – that the brain has imagined. One of imagination's shortcomings, then, is that it takes liberties without telling us it has done so. But if imagination can be too liberal, it can also be too conservative, and that shortcoming has a story of its own.

PART IV

Presentism

presentism (pre·zĕn·tizm)
The tendency for current experience
to influence one's views of the past and the future.

CHAPTER 6

The Future Is Now

Thy letters have transported me beyond
This ignorant present, and I feel now
The future in the instant.

Shakespeare, *Macbeth*

MOST REASONABLY SIZED LIBRARIES have a shelf of futurist tomes from the 1950s with titles such as *Into the Atomic Age* and *The World of Tomorrow.* If you leaf through a few of them, you quickly notice that each of these books says more about the times in which it was written than about the times it was meant to foretell. Flip a few pages and you'll find a drawing of a housewife with a Donna Reed hairdo and a poodle skirt flitting about her atomic kitchen, waiting for the sound of her husband's rocket car before getting the tuna casserole on the table. Flip a few more and you'll see a sketch of a modern city under a glass dome, complete with nuclear trains, antigravity cars and well-dressed citizens gliding smoothly to work on conveyor-belted sidewalks. You will also notice that some things are missing. The men don't carry babies, the women don't carry briefcases, the children don't have pierced eyebrows or nipples, and the mice go *squeak* instead of *click*. There are no skateboarders or panhandlers, no smartphones or smartdrinks, no spandex, latex, Gore-Tex, Amex, FedEx or Wal-Mart. What's more, all the people of African, Asian and Hispanic origin seem to have missed the future entirely. Indeed, what makes these drawings so charming is that they are utterly, fabulously and ridiculously wrong. How could anyone ever have thought that the future would look like some hybrid of *Forbidden Planet* and *Father Knows Best*?

More of the Same

Underestimating the novelty of the future is a time-honoured tradition. Lord William Thomson Kelvin was one of the most farsighted physicists of the nineteenth century (which is why we measure temperature in kelvins), but when he looked carefully into the world of tomorrow he concluded that 'heavier-than-air flying machines are impossible'.[1] Most of his fellow scientists agreed. As the eminent astronomer Simon Newcomb wrote in 1906: 'The demonstration that no possible combination of known substances, known forms of machinery, and known forms of force, can be united in a practical machine by which man shall fly long distances through the air, seems to the writer as complete as it is possible for the demonstration of any physical fact to be.'[2]

Even Wilbur Wright, who proved Kelvin and Newcomb wrong, admitted that in 1901 he had said to his brother that 'man would not fly for fifty years'.[3] He was off by forty-eight. The number of respected scientists and accomplished inventors who declared the airplane an impossibility is exceeded only by the number who said the same thing about space travel, television sets, microwave ovens, nuclear power, heart transplants and female senators. The litany of faulty forecasts, missed marks and prophetic pratfalls is extensive, but let me ask you to ignore for a moment the sheer number of such mistakes and notice instead the similarity of their forms. The writer Arthur C. Clarke formulated what has come to be known as Clarke's first law: 'When a distinguished but elderly scientist states that something is possible he is almost certainly right. When he states that something is impossible, he is very probably wrong.'[4] In other words, when scientists make erroneous predictions, they almost *always* err by predicting that the future will be too much like the present.

Presentism in the Past

Ordinary people are quite scientific in this regard. We have already seen how brains make ample use of the filling-in trick when they remember the past or imagine the future, and the phrase 'filling in' suggests an image of a hole (for example, in a wall or a tooth) being plugged with some sort of material (Spackle or silver). As it turns out, when brains plug holes in their conceptualizations of yesterday and tomorrow, they tend to use a material called *today*. Consider how often this happens when we try to remember the past. When college students hear persuasive speeches that demonstrably change their political opinions, they tend to remember that they always felt as they currently feel.[5] When dating couples try to recall what they thought about their romantic partners two months earlier, they tend to remember that they felt then as they feel now.[6] When students receive their grades on an exam, they tend to remember being as concerned about the exam before they took it as they currently are.[7] When patients are asked about their headaches, the amount of pain they are feeling at the moment determines how much pain they remember feeling the previous day.[8] When middle-aged people are asked to remember what they thought about premarital sex, how they felt about political issues or how much alcohol they drank when they were in college, their memories are influenced by how they think, feel and drink now.[9] When widows and widowers are asked how much grief they felt when their spouse died five years earlier, their memories are influenced by the amount of grief they currently feel.[10] The list goes on, but what's important to notice for our purposes is that in each of these instances, people misremember their own pasts by recalling that they once thought, did and said what they now think, do and say.[11]

This tendency to fill in the holes in our memories of the past with material from the present is especially powerful when it comes to remembering our emotions. In 1992, after announcing on a syndicated television talk show that he would like to live in the White House, Ross Perot became the overnight messiah of a disaffected electorate. For the first time in American history it looked as though a man who had never held office and was not the nominee of a major political party might well win the most powerful job on earth. His supporters were enthusiastic and optimistic. But on 16 July 1992, as suddenly as he had burst onto the scene, Perot withdrew from the race, citing vague concerns about political 'dirty tricks' that might spoil his daughter's wedding. His supporters were devastated. Then, in October of the same year, Perot had yet another change of heart and reentered the race, which he ultimately lost the next month. Between his initial surprising announcement, his even more surprising withdrawal, his unbelievably surprising reentry and his unsurprising defeat, those who supported him experienced a variety of intense emotions. Fortunately, a researcher was on hand to measure these emotional reactions in July, after Perot's withdrawal, and then again in November, after his defeat.[12] The researcher also asked volunteers in November to recall how they had felt in July, and the findings were striking. Those who remained loyal to Perot throughout his flipping and flopping remembered feeling less sad and angry when he withdrew in July than they actually had been, whereas those who abandoned him when he abandoned them remembered being less hopeful than they had been. In other words, Perot supporters erroneously recalled feeling about Perot then as they felt about him now.

Presentism in the Future

If the past is a wall with some holes, the future is a hole with no walls. Memory *uses* the filling-in trick, but imagination *is* the filling-in trick, and if the present lightly colours our remembered pasts, it thoroughly infuses our imagined futures. More simply said, most of us have a tough time imagining a tomorrow that is terribly different from today, and we find it particularly difficult to imagine that we will ever think, want or feel differently than we do now.[13] Teenagers get tattoos because they are confident that DEATH ROCKS will always be an appealing motto, new mothers abandon promising law careers because they are confident that being at home with their children will always be a rewarding job, and smokers who have just finished a cigarette are confident for at least five minutes that they can easily quit and that their resolve will not diminish with the nicotine in their bloodstreams. Psychologists have nothing on teenagers, smokers and mothers. I can recall a Thanksgiving (well, actually, most Thanksgivings) when I ate so much that I realized only as I swallowed my last bite of pumpkin pie that my breathing had become shallow and laboured because my lungs no longer had room to expand. I staggered to the living room, fell flat on the couch, and, as I descended mercifully into a tryptophan coma, was heard to utter these words: 'I'll never eat again.' But, of course, I did eat again – possibly that night, surely within twenty-four hours and probably turkey. I suppose I knew that my vow was absurd even as I made it, and yet, some part of me seemed sincerely to believe that chewing and swallowing were nasty habits that I could easily renounce, if only because the torpid mass that was winding its way through my digestive tract at the approximate speed of continental drift would supply all my nutritional, intellectual and spiritual needs forevermore.

I am appropriately embarrassed by this incident on several counts. First, I ate like a pig. Second, although I had eaten like a pig before and should therefore have known from experience that pigs always end up back at the trough, I really did think that *this time* I might not eat again for days, maybe weeks, maybe ever. I take small comfort in the fact that other pigs seem susceptible to precisely the same delusion. Research in laboratories and supermarkets has demonstrated that when people who have recently eaten try to decide what they will want to eat next week, they reliably underestimate the extent of their future appetites.[14] The double-thick milkshakes, chicken-salad sandwiches and jalapeño sausage pockets that they recently slurped, snarfed and swallowed do not temporarily lower their intelligence. Rather, these people just find it difficult to imagine being hungry when they are full and thus can't bring themselves to provide adequately for hunger's inevitable return. We go shopping after a full breakfast of eggs, toast and bacon, end up buying too few groceries, and then, when the urge for double chocolate chip ice cream makes its regular nightly visit, we curse ourselves for having shopped so lightly.

What is true of sated stomachs is also true of sated minds. In one study, researchers challenged some volunteers to answer five geography questions and told them that after they had taken their best guesses they would receive one of two rewards: either they would learn the correct answers to the questions they had been asked and thus find out whether they had gotten them right or wrong, or they would receive a chocolate bar but never learn the answers.[15] Some volunteers chose their reward *before* they took the geography quiz, and some volunteers chose their reward only *after* they took the quiz. As you might expect, people preferred the chocolate bar before taking the quiz, but they preferred the answers after taking the quiz. In other words, taking the quiz made people so curious that they valued the answers more than a scrumptious piece of chocolate. But do people know this will happen? When a new group of volunteers was asked to *predict* which reward they would choose before and after taking the quiz, these volunteers predicted that they would choose the chocolate bar in both cases. These volunteers – who had not actually *experienced* the intense curiosity that taking the quiz produced – simply couldn't imagine that they would ever forsake a Snickers for a few dull facts about cities and rivers. This finding brings to mind that wonderful scene in the 1967 film *Bedazzled* in which the devil spends his days in bookstores, ripping the final pages out of the mystery novels. This may not strike you as an act so utterly evil that it would warrant Lucifer's personal attention, but when you arrive at the end of a good whodunit only to find the whodunit part missing, you understand why people might willingly trade their immortal souls for the dénouement. Curiosity is a powerful urge, but when you aren't smack-dab in the middle of feeling it, it's hard to imagine just how far and fast it can drive you.

These problems with forecasting our hungers – whether gustatory, sexual, emotional, social or intellectual – are all too familiar. But why? Why are the powers of human imagination so easily humbled? This is, after all, the same imagination that produced space travel, gene therapy, the theory of relativity and the Monty Python cheese-shop sketch. Even the least imaginative among us can imagine things so wild and weird that our mothers would wash our heads out with soap if only they knew. We can imagine being elected to Congress, dropped from a helicopter, painted purple and rolled in almonds. We can imagine life on a banana plantation and inside a submarine. We can imagine being slaves, warriors, sheriffs, cannibals, courtesans, scuba divers and tax collectors. And yet, for some reason, when our bellies are stuffed with mashed potatoes and cranberry sauce, we can't imagine being *hungry*? How come?

Sneak Prefeel

The answer to this question takes us deep into the nature of imagination itself. When we imagine objects, such as penguins, paddleboats or Scotch-tape dispensers, most of us have the experience of actually *seeing* a somewhat sketchy picture of the object in our heads. If I were to ask you whether a penguin's flippers are longer or shorter than its feet, you would probably have the sense of conjuring up a mental image from airy nothing and then 'looking' at it to determine the answer. You would feel as though a picture of a penguin just popped into your head because you wanted it to, and you would have the sense of staring at the flippers for a moment, looking down and checking out the feet, glancing back up at the flippers and then giving me an answer. What you were doing would feel a lot like seeing because, in fact, it is. The region of your brain that is normally activated when you see objects with your eyes – a sensory area called the visual cortex – is also activated when you inspect mental images with your mind's eye.[16] The same is true of other senses. For instance, if I were to ask on which syllable the high note in 'Happy Birthday' is sung, you would probably play the melody in your imagination and then 'listen' to it to determine where the pitch rises and falls. Again, this sense of 'listening with your mind's ear' is not just a figure of speech (especially since no one actually says this). When people imagine sounds, they show activation in a sensory area of the brain called the auditory cortex, which is normally activated only when we hear real sounds with our ears.[17]

These findings tell us something important about how the brain imagines, namely, that it enlists the aid of its sensory areas when it wants to imagine the sensible features of the world. If we want to know how a particular object looks when the object isn't sitting there in front of us, we send information about the object from our memory to our visual cortex, and we experience a mental image. Similarly, if we want to know how a melody sounds when it isn't currently on the radio, we send information about the object from our memory to our auditory cortex, and we experience a mental sound. Because penguins live in Antarctica and 'Happy Birthday' is sung only on birthdays, neither of these things is usually there when we want to inspect it. When our eyes and ears do not feed the visual and auditory cortices the information they require to answer the questions we are asked, we request that the information be sent from memory, which allows us to take a fake look and have a fake listen. Because our brains can do this trick, we are able to discover things about songs (the high note occurs on *birth*) and birds (the flippers are longer than the feet) even when we are all alone in a closet.

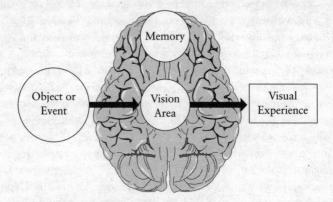

Fig. 10. Visual perception *(above)* gets information from objects and events in the world, whereas visual imagination *(below)* gets information from memory.

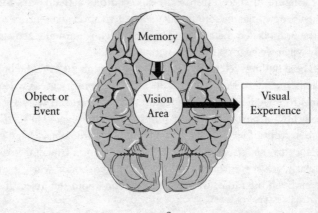

Using the visual and auditory areas to execute acts of imagination is a truly ingenious bit of engineering, and evolution deserves the Microsoft Windows Award for installing it in every one of us without asking permission. But what do seeing and hearing have to do with Thanksgiving gluttons like us – well, at least like me? As it turns out, the imaginative processes that allow us to discover how a penguin looks even when we are locked in a closet are the same processes that allow us to discover how the future will feel when we are locked in the present. The moment someone asks you how much you would enjoy finding your partner in bed with the postman, you *feel* something. Probably something not so good. Just as you generate a mental image of a penguin and then visually inspect it in order to answer questions about its flippers, so do you generate a mental image of an infidelity and then emotionally react to it in order to answer questions about your future feelings.[18] The areas of your brain that respond emotionally to real events respond emotionally to imaginary events as well, which is why your pupils probably dilated and your blood pressure probably rose when I asked you to imagine this particular instance of special delivery.[19] This is a clever method for predicting future feelings, because how we feel when we imagine an event is usually a good indicator of how we will feel when the event itself transpires. If mental images of rapid breathing and flailing mailbags induce pangs of jealousy and waves of anger, then we should expect a real infidelity to do so with even greater swiftness and reliability.

It doesn't take something as emotionally charged as infidelity to illustrate this fact. Every day we say things like 'Pizza sounds pretty good to me', and despite the literal meaning of that utterance, we are not commenting on the acoustic properties of mozzarella. Rather, we are saying that when we imagine eating pizza we experience a small, lovely feeling, and that we interpret this feeling as an indicator of the even larger and lovelier feeling we would experience if we could just get the pizza out of our imaginations and into our mouths. When a Chinese host offers us an appetizer of sautéed spider or crispy grasshopper, we don't have to chew one to know how much we'd dislike the actual experience, because the mere thought of eating bugs causes most Westerners to shudder in disgust, and that shudder tells us that the real thing is likely to induce full-blown nausea. The point here is that we generally do not sit down with a sheet of paper and start logically listing the pros and cons of the future events we are contemplating, but rather, we contemplate them by simulating those events in our imaginations and then noting our emotional reactions to that simulation. Just as imagination *previews* objects, so does it *prefeel* events.[20]

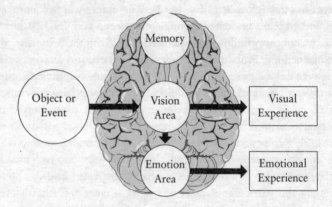

Fig. 11. Both feeling *(above)* and prefeeling *(below)* get information from the vision area, but the vision area gets information from different sources.

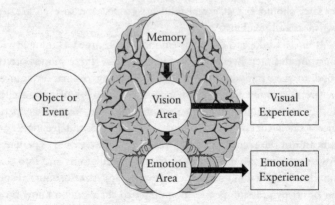

The Power of Prefeeling

Prefeeling often allows us to predict our emotions better than logical thinking does. In one study, researchers offered volunteers a reproduction of an Impressionist painting or a humorous poster of a cartoon cat.[21] Before making their choices, some volunteers were asked to think logically about why they thought they might like or dislike each poster (thinkers), whereas others were encouraged to make their choices quickly and 'from the gut' (nonthinkers). Career counsellors and financial advisors always tell us that we should think long and hard if we wish to make sound decisions, but when the researchers phoned the volunteers later and asked how much they liked their new objet d'art, the thinkers were the least satisfied. Rather than choosing the poster that had made them feel happy when they imagined hanging it in their homes, thinkers had ignored their prefeelings and had instead chosen posters that possessed the qualities of which a career counsellor or financial advisor would approve ('The olive green in the Monet may clash with the curtains, whereas the Garfield poster will signal to visitors that I have a scintillating sense of humour'). Nonthinkers, on the other hand, trusted their prefeelings: they imagined the poster on their wall, noted how they felt when they did so, and assumed that if imagining the poster on their wall made them feel good, then actually seeing it on their wall would probably do the same. And they were right. Prefeeling allowed nonthinkers to predict their future satisfaction more accurately than thinkers did. Indeed, when people are prevented from feeling emotion in the present, they become temporarily unable to predict how they will feel in the future.[22]

But prefeeling has limits. How we feel when we imagine something is not *always* a good guide to how we will feel when we see, hear, wear, own, drive, eat or kiss it. For example, why do you close your eyes when you want to visualize an object, or jam your fingers in your ears when you want to remember the melody of a certain song? You do these things because your brain must use its visual and auditory cortices to execute acts of visual and auditory imagination, and if these areas are already busy doing their primary jobs –

namely, seeing and hearing things in the real world – then they are not available for acts of imagination.[23] You cannot easily imagine a penguin when you are busy inspecting an ostrich because vision is already using the parts of your brain that imagination needs. Put differently, when we ask our brains to look at a real object and an imaginary object at the same time, our brains typically grant the first request and turn down the second. The brain considers the perception of reality to be its first and foremost duty, thus your request to borrow the visual cortex for a moment is expressly and summarily denied. If the brain didn't have this Reality First policy, you'd drive right through a red light if you just so happened to be thinking about a green one. The policy that makes it difficult to imagine penguins when we are looking at ostriches also makes it difficult to imagine lust when we are feeling disgust, affection when we are feeling anger, or hunger when we are feeling full. If a friend were to wreck your new car and then offer to make amends by taking you to a baseball game the following week, your brain would be too busy responding to the car wreck to simulate your emotional response to the game. Future events may request access to the emotional areas of our brains, but current events almost always get the right of way.

The Limits of Prefeeling

We can't see or feel two things at once, and the brain has strict priorities about what it will see, hear and feel and what it will ignore. Imagination's requests are often denied. Both the sensory and emotional systems enforce this policy, and yet, we seem to recognize when the sensory systems are turning down imagination's requests but fail to recognize when the emotional system is doing the same. For instance, if we try to imagine a penguin while we are looking at an ostrich, the brain's policy won't allow it. We understand this, and thus we never become confused and mistakenly conclude that the large bird with the long neck that we are currently seeing is, in fact, the penguin that we were attempting to imagine. The visual experience that results from a flow of information that originates in the world is called *vision*; the visual experience that results from a flow of information that originates in memory is called *mental imagery*;

and while both kinds of experience are produced in the visual cortex, it takes a great deal of vodka before we mix them up.[24] One of the hallmarks of a visual experience is that we can almost always tell whether it is the product of a real or an imagined object. But not so with emotional experience. The emotional experience that results from a flow of information that originates in the world is called *feeling*; the emotional experience that results from a flow of information that originates in memory is called *prefeeling*; and mixing them up is one of the world's most popular sports.

For example, in one study, researchers telephoned people in different parts of the country and asked them how satisfied they were with their lives.[25] When people who lived in cities that happened to be having nice weather that day imagined their lives, they reported that their lives were relatively happy; but when people who lived in cities that happened to be having bad weather that day imagined their lives, they reported that their lives were relatively unhappy. These people tried to answer the researcher's question by imagining their lives and then asking themselves how they felt when they did so. Their brains enforced the Reality First policy and insisted on reacting to real weather instead of imaginary lives. But apparently, these people didn't *know* their brains were doing this and thus they mistook reality-induced feelings for imagination-induced prefeelings.

In a related study, researchers asked people who were working out at a local gym to predict how they would feel if they became lost while hiking and had to spend the night in the woods with neither food nor water.[26] Specifically, they were asked to predict whether their hunger or their thirst would be more unpleasant. Some people made this prediction just after they had worked out on a treadmill (thirsty group), and some made this prediction before they worked out on a treadmill (nonthirsty group). The results showed that 92 percent of the people in the thirsty group predicted that if they were lost in the woods, thirst would be more unpleasant than hunger, but only 61 percent of the people in the nonthirsty group made that prediction. Apparently, the thirsty people tried to answer the researcher's question by imagining being lost in the woods without food and water and then asking themselves how they felt when they did so. But their brains enforced the Reality First policy and insisted on reacting to the real workout rather than the imaginary hike. Because these people didn't *know* their brains were doing this, they confused their feelings and prefeelings.

You've probably been in a similar conundrum yourself. You've had an awful day – the cat peed on the rug, the dog peed on the cat, the washing machine is broken, *World Wrestling* has been pre-empted by *Masterpiece Theatre* – and you naturally feel out of sorts. If at that moment you try to imagine how much you would enjoy playing cards with your buddies the next evening, you may mistakenly attribute feelings that are due to the misbehaviour of real pets and real appliances ('I feel annoyed') to your imaginary companions ('I don't think I'll go because Nick always ticks me off'). Indeed, one of the hallmarks of depression is that when depressed people think about future events, they cannot imagine liking them very much.[27] *Holiday? Romance? A night on the town? No thanks, I'll just sit here in the dark*. Their friends get tired of seeing them flail about in a thick blue funk, and they tell them that this too shall pass, that it is always darkest before the dawn, that every dog has its day and several other important clichés. But from the depressed person's point of view, all the flailing makes perfectly good sense because when they imagine the future, they find it difficult to feel happy today and thus difficult to believe that they will feel happy tomorrow.

We cannot feel good about an imaginary future when we are busy feeling bad about an actual present. But rather than recognizing that this is the inevitable result of the Reality First policy, we mistakenly assume that the future event is the *cause* of the unhappiness we feel when we think about it. Our confusion seems terribly obvious to those who are standing on the sidelines, saying things like 'You're feeling low right now because Pa got drunk and fell off the porch, Ma went to jail for whupping Pa, and your pickup truck got repossessed – but everything will seem different next week and you'll really wish you'd decided to go with us to the opera.' At some level we recognize that our friends are probably right. Nonetheless, when we try to overlook, ignore or set aside our current gloomy state and make a forecast about how we will feel tomorrow, we find that it's a lot like trying to imagine the taste of marshmallow while chewing liver.[28] It is only natural that we should imagine the future and then consider how doing so makes us feel, but because our brains are hell-bent on responding to current events, we mistakenly conclude that we will feel tomorrow as we feel today.

Onward

I've been waiting a long, long time to show someone this cartoon (figure 12), which I clipped from a newspaper in 1983 and have kept tacked to one bulletin board or another ever since. It never fails to delight me. The sponge is being asked to imagine without limits – to envision what it might like to be if the entire universe of possibilities were open to it – and the most exotic thing it can imagine becoming is an arthropod. The cartoonist isn't making fun of sponges, of course; he's making fun of us. Each of us is trapped in a place, a time and a circumstance, and our attempts to use our minds to transcend those boundaries are, more often than not, ineffective. Like the sponge, we think we are thinking outside the box only because we can't see how big the box really is. Imagination cannot easily transcend the boundaries of the present, and one reason for this is that it must borrow machinery that is owned by perception. The fact that these two processes must run on the same platform means that we are sometimes confused about which one is running. We assume that what we feel as we imagine the future is what we'll feel when we get there, but in fact, what we feel as we imagine the future is often a response to what's happening in the present. The time-share

Fig. 12.

arrangement between perception and imagination is one of the causes of presentism, but it is not the only one. So if the train hasn't yet arrived at your stop, if you aren't quite ready to turn out the light and go to sleep, or if the people at Starbucks aren't giving you dirty looks as they get out the mops, let's explore another.

CHAPTER 7

Time Bombs

'And yet not cloy thy lips with loathed satiety,
But rather famish them amid their plenty,
Making them red and pale with fresh variety –
Ten kisses short as one, one long as twenty.
A summer's day will seem an hour but short,
Being wasted in such time-beguiling sport.'

Shakespeare, *Venus and Adonis*

NO ONE HAS EVER WITNESSED the passage of a flying Winnebago, and everyone has witnessed the passage of time. So why is it so much easier to imagine the former than the latter? Because as unlikely as it is that a twenty-thousand-pound recreational vehicle could ever achieve sufficient lift to become airborne, a flying Winnebago would at least look like *something*, and thus we have no trouble producing a mental image of one. Our extraordinary talent for creating mental images of concrete objects is one of the reasons why we function so effectively in the physical world.[1] If you imagine a grapefruit sitting atop a round oatmeal box and then imagine tilting the box away from you, you can actually preview the grapefruit as it falls, and you can see that it will fall toward you when you tilt the box quickly but fall away from you when you tilt the box slowly. Such acts of imagination allow you to reason about the things you are imagining and hence solve important problems in the real world, such as how to get a grapefruit into your lap when you really need one. But time is no grapefruit. It has no colour, shape, size or texture. It cannot be poked, peeled, prodded, pushed, painted or pierced. Time is not an object but an abstraction, hence it does not lend itself to imagery, which is why filmmakers are forced to represent the passage of time with contrivances that involve visible objects, such as calendar leaves blowing in the wind or clocks spinning at warp speed. And yet, predicting our emotional futures requires that we think in and about and across swathes of time. If we can't create a mental image of an abstract concept such as time, then how do we think and reason about it?

SpaceThink

When people need to reason about something abstract, they tend to imagine something concrete that the abstract thing is *like* and then reason about that instead.[2] For most of us, *space* is the concrete thing that *time* is like.[3] Studies reveal that people all over the world imagine time as though it were a spatial dimension, which is why we say that the past is *behind* us and the future is *in front* of us, that we are *moving toward* our senescence and *looking back* on our infancy, and that days *pass us by* in much the same way that a flying Winnebago might. We think and speak as though we were actually *moving away* from a yesterday that is located *over there* and *toward* a tomorrow that is located 180 degrees about. When we draw a time line, those of us who speak English put the past on the left, those of us who speak Arabic put the past on the right,[4] and those of us who speak Mandarin put the past on the bottom.[5] But regardless of our native tongue, we all put the past *someplace* – and the future *someplace else*. Indeed, when we want to solve a problem that involves time – for instance, 'If I ate breakfast before I walked the dog but after I read the newspaper, then what did I do first?' – most of us imagine putting three objects (breakfast, dog, newspaper) in an orderly line and then checking to see which one is furthest to the left (or right, or bottom, depending on our language). Reasoning by metaphor is an ingenious technique that allows us to remedy our weaknesses by capitalizing on our strengths – using things we can visualize to think, talk and reason about things we can't.

Alas, metaphors can mislead as well as illuminate, and our tendency to imagine time as a spatial dimension does both of these things. For example, imagine that you and a friend have managed to get a table at a chic new restaurant with a three-month waiting list, and that after browsing the menus you have discovered that you both want the wasabi-encrusted partridge. Now, each of you has sufficient social grace to recognize that placing identical orders at a fine restaurant is roughly equivalent to wearing matching mouse ears in the main dining room, so you decide instead that one of you will order the partridge, the other will order the venison gumbo, and that you will then share them oh so fashionably. You do this not only to avoid being mistaken for tourists but also because you believe that variety is the spice of life. There are very few homilies involving spices, and this one is as good as they get. Indeed, if we were to measure your pleasure after the meal, we would probably find that you and your friend are happier with the sharing arrangement than either of you would have been had you each had a full order of partridge to yourselves.

But something strange happens when we extend this problem in time. Imagine that the maître d' is so impressed by your sophisticated ensemble that he invites you (but alas, not your friend, who really could use a new look) to return on the first Monday of every month for the next year to enjoy a free meal at his best table. Because the kitchen occasionally runs short of ingredients, he asks you to decide right now what you would like to eat on each of your return visits so that he can be fully prepared to pamper you in the style to which you are quickly becoming accustomed. You flip back through the menu. You hate rabbit, veal is politically incorrect, you are appropriately apathetic about vegetable lasagna, and as you scan the list you decide that there are just four dishes that strike your rapidly swelling fancy: the partridge, the venison gumbo, the blackened sea bream and the saffron seafood risotto. The partridge is clearly your favourite, and even without a pear tree you are tempted to order twelve of them. But that would be so gauche, so déclassé, and what's more, you would miss the spice of life. So you ask the maître d' to prepare the partridge every other month, and to fill in the remaining six meals with equal episodes of gumbo, sea bream and risotto.

You may be one snappy dresser, *mon ami*, but when it comes to food, you have just cooked your own goose.[6] Researchers studied this experience by inviting volunteers to come to the laboratory for a snack once a week for several weeks.[7] They asked some of the volunteers (choosers) to choose all their snacks in advance, and – just as you did – the choosers usually opted for a healthy dose of variety. Next, the researchers asked a new group of volunteers to come to the lab once a week for several weeks. They fed some of these volunteers their favourite snack every time (no-variety group), and they fed other volunteers their favourite snack on most occasions and their second-favourite snack on others (variety group). When they measured the volunteers' satisfaction over the course of the study, they found that volunteers in the no-variety group were more satisfied than were volunteers in the variety group. In other words, variety made people *less* happy, not *more*. Now wait a second – there's something fishy here, and it isn't the sea bream. How can variety be the spice of life when one sits down with a friend at a fancy restaurant but the bane of one's existence when one orders snacks to be consumed in successive weeks?

Among life's cruellest truths is this one: wonderful things are especially wonderful the first time they happen, but their wonderfulness wanes with repetition.[8] Just compare the first and last time your child said 'Mama' or your partner said 'I love you' and you'll know exactly what I mean. When we have an experience – hearing a particular sonata, making love with a particular person, watching the sun set from a particular window of a particular room – on successive occasions, we quickly begin to adapt to it, and the experience yields less pleasure each time. Psychologists call this *habituation*, economists call it *declining marginal utility*, and the rest of us call it marriage. But human beings have discovered two devices that allow them to combat this tendency: variety and time. One way to beat habituation is to increase the variety of one's experiences ('Hey, honey, I have a kinky idea – let's watch the sun set from *the kitchen* this time').[9] Another way to beat habituation is to increase the amount of time that separates repetitions of the experience. Clinking champagne glasses and kissing one's spouse at the stroke of midnight would be a relatively dull exercise were it to happen every evening, but if one does it on New Year's Eve and then allows a full year to pass before doing it again, the experience will offer an endless bouquet of delights because a year is plenty long enough for the effects of habituation to disappear. The point here is that time and variety are two ways to avoid habituation, and if you have one, then you don't need the other. In fact when episodes are sufficiently separated in time, variety is not only unnecessary – it can actually be costly.

I can illustrate this fact with some precision if you allow me to make a few reasonable assumptions. First, imagine that we can use a machine called a hedonimeter to measure a person's pleasure in *hedons*. Let's start by making a *favouring assumption*: let's assume that the first bite of partridge provides you with, say, fifty hedons, whereas the first bite of gumbo provides you with forty hedons. This is what it means to say that you favour partridge over gumbo. Second, let's make a *habituation-rate assumption*: let's assume that once you take a bite of a dish, each subsequent bite of the same dish taken within, say, ten minutes, provides one less hedon than the bite before it did. Finally, let's make a *consumption-rate assumption*: let's assume that you normally eat at the brisk pace of one bite every thirty seconds. Figure 13 shows you what happens to your pleasure if we make these assumptions about favouring, habituation rate and consumption rate. As you can see, the best way to maximize your pleasure in this case is to start with the partridge and then switch to gumbo after taking ten bites (which happens after five minutes). Why switch? Because, as the lines show, the eleventh bite of partridge (taken at minute 5.5) would bring you a mere thirty-nine hedons, whereas a bite of the as-yet-untasted gumbo would yield forty. So this is the precise point in the meal at which you and your friend should trade plates, seats or at least mouse ears.[10] But now look at figure 14 and notice how radically things change when we extend this gastronomic episode in time by altering your consumption rate. When your bites are separated by anything greater than ten minutes (in this case, fifteen minutes), then habituation no longer occurs, which means that every bite is as good as the first and a bite of gumbo is *never* better than a bite of partridge. In other words, if you could eat slowly enough, then variety would not only be *unnecessary*, it would actually be *costly*, because a bite of gumbo would *always* provide less pleasure than yet another bite of partridge.

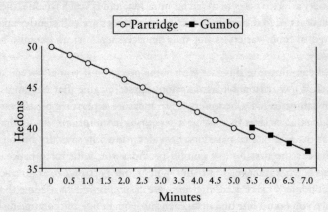

Fig. 13. Variety increases pleasure when consumption is rapid.

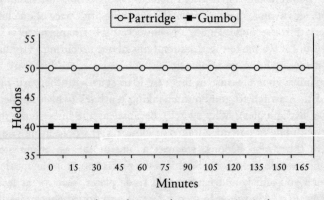

Fig. 14. Variety reduces pleasure when consumption is slow.

Now, when you and your friend sat down at the imaginary restaurant together, you ordered two dishes to be eaten simultaneously. You knew that you wouldn't have much time between bites, so you asked for variety to spice things up. Good call. But when the maître d' asked you to order a sequence of meals in advance, you asked for variety then too. Why did you ask for variety when you already had time? Blame the spatial metaphor (see figure 15). Because you thought of dishes that were separated in time by imagining dishes that were separated by a few inches on a single table, you assumed that what was true of spatially separated dishes would be true of temporally separated dishes as well. When dishes are separated by space, it makes perfectly good sense to seek variety. After all, who would want to sit down at a table with twelve identical servings of partridge? We love sampler plates, pupu platters and all-you-can-eat buffets because we want – and *should* want – variety among alternatives that we will experience in a single episode. The problem is that when we reason by metaphor and think of a dozen successive meals in a dozen successive months as though they were a dozen dishes arranged on a long table in front of us, we mistakenly treat *sequential* alternatives as though they were *simultaneous* alternatives. This is a mistake because sequential alternatives already have time on their side, hence variety makes them less pleasurable rather than more.

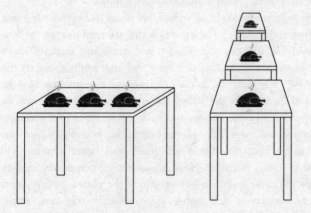

Fig. 15. Simultaneous consumption *(left)* and sequential consumption *(right)*.

Starting Now

Because time is so difficult to imagine, we sometimes imagine it as a spatial dimension. And sometimes we just don't imagine it at all. For example, when we imagine future events, our mental images will usually include the relevant people, places, words and actions, but they rarely include a clear indication of the *time* at which those people in those places are speaking and acting. When we imagine ourselves discovering our spouse's infidelity on New Year's Eve, our mental image *looks* very much like a mental image of ourselves discovering the infidelity on Purim, Halloween or Russian Orthodox Easter. Indeed, the mental image of *finding your spouse in bed with the postman on New Year's Eve* changes dramatically when you substitute *barber* for *spouse*, or *conversation* for *bed*, but hardly at all when you substitute *Thanksgiving* for *New Year's Eve*. In fact, it is just about impossible to make this substitution because, alas, there is nothing in the mental image to change. We can inspect a mental image and see *who* is doing *what* and *where*, but not *when* they are doing it. In general, mental images are *atemporal*.[11]

So how do we decide how we will feel about things that are going to happen in the future? The answer is that we tend to imagine how we would feel if those things happened *now*, and then we make some allowance for the fact that *now* and *later* are not exactly the same thing. For instance, ask a heterosexual teenage boy how he would feel if one of the Budweiser babes were to show up at his door *right now*, bikini-clad, cooing and in desperate need of a massage. His reaction will be visible. He will smile, his eyes will widen, his pupils will dilate, his cheeks will flush and other systems will respond as nature intended. Now, if you ask a different teenage boy precisely the same question but substitute the phrase *in fifty years* for *right now*, you will notice approximately the same initial response. Indeed, for a moment you might even suspect that this second teenager is focusing entirely on his mental image of the barefoot goddess with the bee-sting lips and that he is failing to consider the fact that this imaginary event is supposed to be taking place a half-century hence. But give him some time – say, a few hundred milliseconds. As the milliseconds pass, you will notice that his initial rush of enthusiasm fades as he considers the date of the imaginary event, realizes that adolescent males have one set of needs and grandfathers another, and correctly concludes that a cameo appearance by a nubile nymphet will probably not be quite as stimulating in his golden years as it would be in his testosterone-charged present. His initial flip and subsequent flop are quite revealing because they suggest that when he was asked to imagine the future event, he began by imagining the event as though it were happening in the present and only *then* considered the fact that the event would take place in the future, when maturity will have taken its inevitable toll on his eyesight and his libido.

Why does this matter? After all, in the final analysis the teenager did take into account the fact that *now* and *five decades from now* are not the same thing, so who cares if he considered this fact only *after* he was momentarily transfixed by his mental image of the airbrushed vixen from Planet Bud? I care. And you should care too. By imagining the event happening *now* and then correcting for the fact that it was actually going to happen *later*, the teenager used a method for making judgments that is quite common but that inevitably leads to error.[12] To understand the nature of this error, consider a study in which volunteers were asked to guess how many African countries belonged to the United Nations.[13] Rather than answering the question straightaway, the volunteers were asked to make their judgments by using the flip-then-flop method. Some volunteers were asked to give their answer by saying how much larger or smaller it was than ten, and other volunteers were asked to answer by saying how much larger or smaller it was than sixty. In other words, volunteers were given an arbitrary *starting point* and were asked to *correct* it until they reached an appropriate *ending point* – just as the teenager used an image of a beautiful woman in the present moment as a starting point for his judgment ('I'm wildly excited!') and then corrected it to achieve an ending point for his judgment ('But since I'll be sixty-seven years old when all this happens, I probably won't be quite as excited as I am now').

The problem with this method of making judgments is that starting points have a profound impact on ending points. Volunteers who started with ten guessed that there were about twenty-five African nations in the UN, whereas volunteers who started with sixty guessed that there were about forty-five. Why such different answers? Because volunteers began their task by asking themselves whether the starting point could be the right answer, and then, realizing that it could not, moved slowly toward a more reasonable one ('Ten can't be right. How about twelve? No, still too low. Fourteen? Maybe twenty-five?').[14] Alas, because this process requires time and attention, the group that started with ten and the group that started with sixty got tired and quit before they met in the middle. This really isn't so strange. If you asked a child to count upward from zero and another child to count downward from a million, you could be pretty sure that when they finally got exhausted, gave up and went off in search of eggs to throw at your garage door, they would have reached very different numbers. Starting points matter because we often end up close to where we started.

When people predict future feelings by imagining a future event as though it were happening in the present and then correcting for the event's actual location in time, they make the same error. For example, volunteers in one study were asked to predict how much they would enjoy eating a bite of spaghetti bolognese the next morning or the next afternoon.[15] Some of the volunteers were hungry when they made this prediction, and some were not. When volunteers made these predictions under ideal conditions, they predicted that they would enjoy spaghetti more in the afternoon than in the morning, and their current hunger had little impact on their predictions. But some of the volunteers made these predictions under less-than-ideal conditions. Specifically, they were asked to make these predictions while simultaneously performing a second task in which they had to identify musical tones. Research has shown that performing a simultaneous task such as this one causes people to stay very close to their starting points. And indeed, when volunteers made predictions while identifying musical tones, they predicted that they would like spaghetti just as much in the morning as in the afternoon. What's more, their current hunger had a strong impact on their predictions, so that hungry volunteers expected to like spaghetti the next day (no matter when they ate it) and sated volunteers expected to dislike spaghetti the next day (no matter when they ate it). This pattern of results suggests that all volunteers made their predictions by the flip-then-flop method: they first imagined how much they would enjoy eating the spaghetti in the present ('Yum!' if they were hungry and 'Yuck!' if they were full) and used this prefeeling as a starting point for their prediction of tomorrow's pleasures. Then, just as the hypothetical teenager corrected his judgment when he considered the fact that his current appreciation of a curvaceous coquette would probably be different fifty years later, the volunteers corrected their judgments by considering the time of day at which the spaghetti would be eaten ('Spaghetti for dinner is terrific, but spaghetti for breakfast? Yuck!'). However, volunteers who had made their predictions while identifying musical tones were unable to correct their judgments, and as such, their ending point was quite close to their starting point. Because we naturally use our present feelings as a starting point when we attempt to predict our future feelings, we expect our future to feel a bit more like our present than it actually will.[16]

Next to Nothing

If you have no special talents or intriguing deformities but are still harbouring a secret desire to get into *Guinness World Records*, here's something you might try: march into your boss's office on Monday morning and say, 'I've been with the company for some time, I believe my work has been excellent, and I'd like a fifteen percent pay cut . . . though I could settle for ten if that's all the firm can manage right now.' The Guinness people will be taking careful notes because in the long and often contentious history of labour relations, it is unlikely that anyone has ever before demanded a pay cut. Indeed, people *hate* pay cuts, but research suggests that the reason they hate pay cuts has very little to do with the *pay* part and everything to do with the *cut* part. For instance, when people are asked whether they would prefer to have a job at which they earned $30,000 the first year, $40,000 the second year and $50,000 the third year, or a job at which they earned $60,000 then $50,000 then $40,000, they generally prefer the job with the increasing wages, despite the fact that they would earn less money over the course of the three years.[17] This is quite curious. Why would people be willing to reduce their total income in order to avoid experiencing a cut in pay?

Comparing with the Past

If you've ever fallen asleep one night with the television blaring and been awakened another night by a single footstep, then you already know the answer. The human brain is not particularly sensitive to the absolute magnitude of stimulation, but it is extraordinarily sensitive to differences and changes – that is, to the *relative* magnitude of stimulation. For example, if I blindfolded you and asked you to hold a wooden block in your hand, would you be able to tell if I then placed a pack of gum on top of it? The right answer is 'It depends', and what it depends on is the weight of the block. If the block weighed only an ounce, then you'd immediately notice the 500 percent increase in weight when I added a five-ounce pack of gum. But if the block weighed ten pounds, then you'd never notice the .03 percent increase in weight. There is no answer to the question 'Can people detect five ounces?' because brains do not detect ounces, they detect changes in ounces and differences in ounces, and the same is true for just about every physical property of an object. Our sensitivity to relative rather than absolute magnitudes is not limited to physical properties such as weight, brightness or volume. It extends to subjective properties, such as value, goodness and worth as well.[18] For instance, most of us would be willing to drive across town to save $50 on the purchase of a $100 radio but not on the purchase of a $100,000 automobile because $50 seems like a fortune when we're buying radios ('Wow, Target has the same radio for half off!') but a pittance when we're buying cars ('Like I'm going to schlep across the city just to get this car for one twentieth of a percent less?').[19]

Economists shake their heads at this kind of behavior and will correctly tell you that your bank account contains absolute dollars and not 'percentages off.' If it is worth driving across town to save $50, then it doesn't matter which item you're saving it on because when you spend these dollars on gas and groceries, the dollars won't know where they came from.[20] But these economic arguments fall on deaf ears because human beings don't think in absolute dollars. They think in relative dollars, and fifty is or isn't a lot of dollars depending on what it is relative to (which is why people who don't worry about whether their mutual-fund manager is keeping 0.5 or 0.6 percent of their investment will nonetheless spend hours scouring the Sunday paper for a coupon that gives them 40 percent off a tube of toothpaste). Marketers, politicians and other agents of influence know about our obsession with relative magnitudes and routinely turn it to their own advantage. For instance, one ancient ploy involves asking someone to pay an unrealistically large cost ('Would you come to our Save the Bears meeting next Friday and then join us Saturday for a protest march at the zoo?') before asking them to pay a smaller cost ('Okay then, could you at least contribute five dollars to our organization?'). Studies show that people are much more likely to agree to pay the small cost after having first contemplated the large one, in part because doing so makes the small cost seems so . . . er, bearable.[21]

Because the subjective value of a commodity is relative, it shifts and changes depending on what we compare the commodity to. For instance, every morning on my walk to work I stop at my neighbourhood Starbucks and hand $1.89 to the barista, who then hands me twenty ounces of better-than-average coffee. I have no idea what it costs Starbucks to make this coffee, and I have no idea why they have chosen to charge me this particular amount, but I do know that if I stopped in one morning and found that the price had suddenly jumped to $2.89, I would immediately do one of two things: I would compare the new price to the price I used to pay, conclude that coffee at Starbucks had gotten too damned expensive, and invest in one of those vacuum-sealed travel mugs and start brewing my coffee at home; or I would compare the new price to the price of other things I could buy with the same amount of cash (e.g., two felt-tip markers, a thirty-two-inch branch of artificial bamboo, or 1/100th of the twenty-CD boxed set *The Complete Miles Davis at Montreux*) and conclude that the coffee at Starbucks was a bargain. In theory I could make either of these comparisons, so which one would I actually make?

We both know the answer to that: I'd make the easy one. When I encounter a $2.89 cup of coffee, it's all too easy for me to recall what I paid for coffee the day before and not so easy for me to imagine all the other things I might buy with my money.[22] Because it is so much easier for me to *remember the past* than to *generate new possibilities*, I will tend to compare the present with the past even when I *ought* to be comparing it with the possible. And that is indeed what I *ought* to be doing because it really doesn't *matter* what coffee cost the day before, the week before or at any time during the Hoover administration. Right now I have absolute dollars to spend and the only question I need to answer is how to spend them in order to maximize my satisfaction. If an international bean embargo suddenly caused the price of coffee to skyrocket to $10,000 per cup, then the only question I would need to ask myself is: 'What else can I do with ten thousand dollars, and will it bring me more or less satisfaction than a cup of coffee?' If the answer is 'more', then I should walk away. If the answer is 'less', then I should get a cup of coffee. And an accountant with a whip.

The fact that it is so much easier to remember the past than to generate the possible causes us to make plenty of weird decisions. For instance, people are more likely to purchase a package holiday that has been marked down from $600 to $500 than an identical package that costs $400 but that was on sale the previous day for $300.[23] Because it is easier to compare a holiday package's price with its former price than with the price of other things one might buy, we end up preferring bad deals that have become decent deals to great deals that were once amazing deals. The same tendency leads us to treat commodities that have a 'memorable past' differently from those that don't. For example, imagine that you have a $20 bill and a $20 concert ticket in your wallet, but when you arrive at the concert you realize that you've lost the ticket en route. Would you buy a new one? Most people say no.[24] Now imagine that instead of a $20 bill and a $20 ticket, you have two $20 bills in your wallet, and when you arrive at the concert you realize that you've lost one of the bills en route. Would you buy a concert ticket? Most people say yes. It doesn't take a logician to see that the two examples are identical in all the ways that matter: in both cases you've lost a piece of paper that was valued at $20 (a ticket or a bill), and in both cases you must now decide whether to spend the money that remains in your wallet on a concert. Nonetheless, our stubborn insistence on comparing the present to the past leads us to reason differently about these functionally equivalent cases. When we lose a $20 bill and then contemplate buying a concert ticket for the first time, the concert has no past, hence we correctly compare the cost of seeing the concert with other possibilities ('Should I spend twenty dollars to see the concert, or should I buy some new sharkskin mittens?'). But when we lose a ticket we've previously purchased and contemplate 'replacing it', the concert has a past, and hence we compare the current cost of seeing the concert ($40) with its previous cost ($20) and feel disinclined to see a performance whose price has suddenly doubled.

Comparing with the Possible

We make mistakes when we compare with the past instead of the possible. When we *do* compare with the possible, we still make mistakes. For example, if you're like me, your living room is a mini-warehouse of durable goods ranging from chairs and lamps to stereos and television sets. You probably shopped around a bit before buying these items, and you probably compared the one you ultimately bought with a few alternatives – other lamps in the same catalogue, other chairs on the showroom floor, other stereos on the same shelf, other televisions at the same mall. Rather than deciding *whether* to spend money, you were deciding *how* to spend money, and all the possible ways of spending your money were laid out for you by the nice people who wanted it. These nice people helped you overcome your natural tendency to compare with the past ('Is this television really that much better than my old one?') by making it extremely easy for you to compare with the possible ('When you see them side by side here in the store, the Panasonic has a much sharper picture than the Sony'). Alas, we are all too easily fooled by such side-by-side comparisons, which is why retailers work so hard to ensure that we make them.

For example, people generally don't like to buy the most expensive item in a category, hence retailers can improve their sales by stocking a few *very* expensive items that no one actually buys ('Oh my God, the 1982 Château Haut-Brion Pessac-Léognan sells for five hundred dollars a bottle!') but that make less expensive items seem like a bargain by comparison ("I'll just stick with the sixty-dollar zinfandel").[25] Unscrupulous real estate agents bring buyers to dilapidated dumps that are conveniently located between a massage parlour and a crack house before bringing them to the ordinary homes that they actually hope to sell, because the dumps make the ordinary homes seem extraordinary ('Oh, look, honey, no needles on the lawn!').[26] Our side-by-side comparisons can be influenced by extreme possibilities such as extravagant wines and dilapidated houses, but they can also be influenced by the addition of extra possibilities that are identical to those we are already considering. For example, in one study, physicians read about Medication X and were then asked whether they would prescribe the medication for a patient with osteoarthritis.[27] The physicians clearly considered the medication worthwhile, because only 28 percent chose not to prescribe it. But when another group of physicians was asked whether they would prescribe Medication X or an equally effective Medication Y for a patient with the same disease, 48 percent chose to prescribe nothing. Apparently, adding another equally effective medication to the list of possibilities made it difficult for the physicians to decide between the two medications, thus leading many of them to recommend neither. If you've ever caught yourself saying, 'I'm having such a hard time deciding between these two movies that I think I'll just stay home and watch reruns instead', then you know why physicians made the mistake they did.[28]

One of the most insidious things about side-by-side comparison is that it leads us to pay attention to *any* attribute that distinguishes the possibilities we are comparing.[29] I've probably spent some of the unhappiest hours of my life in stores that I meant to visit for fifteen minutes. I stop at the mall on the way to the picnic, park the car, dash in and expect to reemerge a few minutes later with a nifty little digital camera in my pocket. But when I get to Wacky Bob's Giant Mega Super Really Big World of Cameras, I am confronted by a bewildering panoply of nifty little digital cameras that differ on many attributes. Some of these are attributes that I would have considered even if there had been only one camera in the display case

('This is light enough to fit in my shirt pocket so I can take it any-where'), and some are attributes I would never have thought about had the differences between cameras not been called to my attention ('The Olympus has flash output compensation, but the Nikon doesn't. By the way, what *is* flash output compensation?'). Because side-by-side comparisons cause me to consider *all* the attributes on which the cameras differ, I end up considering attributes that I don't really care about but that just so happen to distinguish one camera from another.[30] For example, what attributes would you care about if you were shopping for a new dictionary? In one study, people were given the opportunity to bid on a dictionary that was in perfect con-dition and that listed ten thousand words, and on average they bid $24.[31] Other people were given the opportunity to bid on a diction-ary with a torn cover that listed twenty thousand words, and on average they bid $20. But when a third group of people was allowed to compare the two dictionaries side by side, they bid $19 for the small intact dictionary and $27 for the large torn dictionary. Appar-ently, people care about the condition of a dictionary's cover, but they care about the number of words it contains only when that attribute is brought to their attention by side-by-side comparison.

Comparing and Presentism

Now let's step back for a moment and ask what all of these facts about comparison mean for our ability to imagine future feelings. The facts are these: *(a)* value is determined by the comparison of one thing with another; *(b)* there is more than one kind of comparison we can make in any given instance; and *(c)* we may value something more highly when we make one kind of comparison than when we make a different kind of comparison. These facts suggest that if we want to predict how something will make us feel in the future, we *must* consider the kind of comparison we will be making in the future and *not* the kind of comparison we happen to be making in the present. Alas, because we make comparisons without even thinking about them ('Man, that coffee has gotten expensive!' or 'I'm not paying double to see this concert'), we rarely consider the fact that the comparisons we are making now may not be the ones we will be making later.[32] For example, volunteers in one study were asked to sit at a table and predict how much they would enjoy eating crisps a few minutes later.[33] Some of the volunteers saw a bag of crisps and a chocolate bar sitting on the table, and others saw a bag of crisps and a tin of sardines sitting on the table. Did these extraneous foods influence the volunteers' predictions? You bet they did. Volunteers naturally compared the crisps with the extraneous food, and they predicted that they'd like eating the crisps *more* when they compared the crisps to the sardines than when they compared the crisps to chocolate. But they were wrong. Because when volunteers actually *ate* the crisps, the sardine tin and the chocolate bar that were sitting on the table had *no influence whatsoever* on their enjoyment of the crisps. After all, when one has a mouthful of crispy, salty, oily, fried potatoes, another food item that just so happens to be sitting there on the table is largely irrelevant – just as the person you *might have* been making love with is largely irrelevant when you are in the middle of making love with someone else. What the volunteers didn't realize was that the comparisons they made as they *imagined* eating a crisp ('Sure, crisps are okay . . . but chocolate is *so* much better') were not the comparisons they would make when they were actually chowing down on one.

Most of us have had similar experiences. We compare the small, elegant speakers with the huge, boxy speakers, notice the acoustical difference, and buy the hulking leviathans. Alas, the acoustical difference is a difference we never notice again, because when we get the monster speakers home we do not compare their sound to the sound of some speaker we listened to a week earlier at the store, but we do compare their awful boxiness to the rest of our sleek, elegant and now-spoiled décor. Or we travel to France, meet a couple from our hometown, and instantly become touring buddies because compared with all those French people who hate us when we don't try to speak their language and hate us more when we do, the hometown couple seems exceptionally warm and interesting. We are delighted to have found these new friends, and we expect to like them just as much in the future as we do today. But when we have them over for dinner a month after returning home, we are surprised to find that our new friends are rather boring and remote compared with our regular friends, and that we actually dislike them enough to qualify for French citizenship. Our mistake was not in touring Paris with a couple of dull homies but in failing to realize that the comparison we were making in the present ('Lisa and Walter are so much nicer than the waiter at Le Grand Colbert') is not the comparison we would be making in the future ('Lisa and Walter aren't nearly as nice as Rebecca and Dan'). The same principle explains why we love new things when we buy them and then stop loving them shortly thereafter. When we start shopping for a new pair of sunglasses, we naturally contrast the hip, stylish ones in the store with the old, outdated ones that are sitting on our noses. So we buy the new ones and stick the old ones in a drawer. But after just a few days of wearing our new sunglasses we stop comparing them with the old pair, and – well, what do you know? The delight that the comparison produced evaporates.

The fact that we make different comparisons at different times – but don't realize that we will do so – helps explain some otherwise puzzling conundrums. For instance, economists and psychologists have shown that people expect losing a dollar to have more impact than gaining a dollar, which is why most of us would refuse a bet that gives us an 85 percent chance of doubling our life savings and a 15 percent chance of losing it.[34] The likely prospect of a big gain just doesn't compensate for the unlikely prospect of a big loss because we think losses are more powerful than equal-sized gains. But whether we think of something as a gain or a loss often depends on the comparisons we are making. For example, how much is a 1993 Mazda Miata worth? According to my insurance company, the correct answer this year is about $2,000. But as the owner of a 1993 Mazda Miata, I can guarantee that if you wanted to buy my sweet little car with all of its adorable dents and mischievous rattles for a mere $2,000, you'd have to pry the keys out of my cold, dead hands. I also guarantee that if you saw my car, you'd think that for $2,000 I should not only give you the car and the keys but that I should throw in a bicycle, a lawn mower and a lifetime subscription to *The Atlantic*. Why would we disagree about the fair value of my car? Because you would be thinking about the transaction as a potential gain ('Compared with how I feel now, how happy will I be if I get this car?') and I would be thinking about it as a potential loss ('Compared with how I feel now, how happy will I be if I lose this car?').[35] I would want to be compensated for what I expected to be a powerful loss, but you would not want to compensate me because you would be expecting a less powerful gain. What you would be failing to realize is that once you owned my car, your frame of reference would shift, you would be making the same comparison that I am now making, and that the car would be worth every penny you paid for it. What I would be failing to realize is that once I didn't own the car, my frame of reference would shift, I would be making the same comparison that you're making now, and that I'd be delighted with the deal because, after all, I'd never *pay* $2,000 for a car that was identical to the one I just sold you. The reason why we disagree on the price and quietly question each other's integrity and parenthood is that neither of us realizes that the kinds of comparisons we are naturally making as buyers and sellers are not the kinds of comparison we will naturally make once we become owners and former owners.[36] In short, the comparisons we make have a profound impact on our feelings, and when we fail to recognize that the comparisons we are making today are not the comparisons we will make tomorrow, we predictably underestimate how differently we will feel in the future.

Onward

Historians use the word *presentism* to describe the tendency to judge historical figures by contemporary standards. As much as we all despise racism and sexism, these isms have only recently been considered moral turpitudes, and thus condemning Thomas Jefferson for keeping slaves or Sigmund Freud for patronizing women is a bit like arresting someone today for having driven without a seat belt in 1923. And yet, the temptation to view the past through the lens of the present is nothing short of overwhelming. As the president of the American Historical Association noted, 'Presentism admits of no ready solution; it turns out to be very difficult to exit from modernity.'[37] The good news is that most of us aren't historians and thus we don't have to worry about finding that particular exit. The bad news is that all of us are futurians, and presentism is an even bigger problem when people look forward rather than backward. Because predictions about the future are made *in* the present, they are inevitably influenced *by* the present. The way we feel right now ('I'm so hungry') and the way we think right now ('The big speakers sound better than the little ones') exert an unusually strong influence on the way we think we'll feel later. Because time is such a slippery concept, we tend to imagine the future as the present with a twist, thus our imagined tomorrows inevitably look like slightly twisted versions of today. The reality of the moment is so palpable and powerful that it holds imagination in a tight orbit from which it never fully escapes. Presentism occurs because we fail to recognize that our future selves won't see the world the way we see it now. As we are about to learn, this fundamental inability to take the perspective of the person to whom the rest of our lives will happen is the most insidious problem a futurian can face.

PART V

Rationalization

ra·tion·al·i·za·tion (rae·shen·ăl·i·zē·shen)
The act of causing something to be or to
seem reasonable.

CHAPTER 8

Paradise Glossed

For there is nothing either good or bad, but thinking makes it so.

Shakespeare, *Hamlet Prince of Denmark*

FORGET YOGA. Forget liposuction. And forget those herbal supplements that promise to improve your memory, enhance your mood, reduce your waistline, restore your hairline, prolong your lovemaking and improve your memory. If you want to be happy and healthy, you should try a new technique that has the power to transform the grumpy, underpaid chump you are now into the deeply fulfilled, enlightened individual you've always hoped to be. If you don't believe me, then just consider the testimony of some people who've tried it:

- 'I am so much better off physically, financially, mentally, and in almost every other way.' *(JW from Texas)*
- 'It was a glorious experience.' *(MB from Louisiana)*
- 'I didn't appreciate others nearly as much as I do now.' *(CR from California)*

Who are these satisfied customers, and what is the miraculous technique they're all talking about? Jim Wright, former Speaker of the United States House of Representatives, made his remark after committing sixty-nine ethics violations and being forced to resign in disgrace. Moreese Bickham, a former inmate, made his remark upon being released from the Louisiana State Penitentiary where he'd served thirty-seven years for defending himself against the Ku Klux Klansmen who'd shot him. And Christopher Reeve, the dashing star of *Superman*, made his remark after an equestrian accident left him paralysed from the neck down, unable to breathe without the help of a ventilator. The moral of the story? If you want to be happy, healthy, wealthy and wise, then skip the vitamin pills and the plastic surgeries and try public humiliation, unjust incarceration or quadriplegia instead.

Uh-huh. Right. Are we really supposed to believe that people who lose their jobs, their freedom and their mobility are somehow *improved* by the tragedies that befall them? If that strikes you as a far-fetched possibility, then you are not alone. For at least a century, psychologists have assumed that terrible events – such as having a loved one die or becoming the victim of a violent crime – must have a powerful, devastating and enduring impact on those who experience them.[1] This assumption has been so deeply embedded in our conventional wisdom that people who *don't* have dire reactions to events such as these are sometimes diagnosed as having a pathological condition known as 'absent grief'. But recent research suggests that the conventional wisdom is wrong, that the absence of grief is quite normal, and that rather than being the fragile flowers that a century of psychologists have made us out to be, most people are surprisingly resilient in the face of trauma. The loss of a parent or spouse is usually sad and often tragic, and it would be perverse to suggest otherwise. But the fact is that while most bereaved people are quite sad for a while, very few become chronically depressed and most experience relatively low levels of relatively short-lived distress.[2] Although more than half the people in the United States will experience a trauma such as rape, physical assault or natural disaster in their lifetimes, only a small fraction will ever develop any posttraumatic pathology or require any professional assistance.[3] As one group of researchers noted, 'Resilience is often the most commonly observed outcome trajectory following exposure to a potentially traumatic event.'[4] Indeed, studies of those who survive major traumas suggest that the vast majority do quite well, and that a significant portion claim that their lives were *enhanced* by the experience.[5]

I know, I know. It sounds suspiciously like the title of a country song, but the fact is that most people do pretty well when things go pretty bad.

If resilience is all around us, then why are statistics such as these so surprising? Why do most of us find it difficult to believe that *we* could ever consider a lifetime behind bars to be 'a glorious experience'[6] or come to see paralysis as 'a unique opportunity' that gave 'a new direction'[7] to our lives? Why do most of us shake our heads in disbelief when an athlete who has been through several gruelling years of chemotherapy tells us that 'I wouldn't change anything',[8] or when a musician who has become permanently disabled says, 'If I had it to do all over again, I would want it to happen the same way',[9] or when quadriplegics and paraplegics tell us that they are pretty much as happy as everyone else?[10] The claims made by people who have experienced events such as these seem frankly outlandish to those of us who are merely imagining those events – and yet, who are we to argue with the people who've actually been there?

The fact is that negative events do affect us, but they generally don't affect us as much or for as long as we expect them to.[11] When people are asked to predict how they'll feel if they lose a job or a romantic partner, if their candidate loses an important election or their team loses an important game, if they mess up an interview, flunk an exam or fail a contest, they consistently overestimate how awful they'll feel and how long they'll feel awful.[12] Able-bodied people are willing to pay far more to avoid becoming disabled than disabled people are willing to pay to become able-bodied again because able-bodied people underestimate how happy disabled people are.[13] As one group of researchers noted, 'Chronically ill and disabled patients generally rate the value of their lives in a given health state more highly than do hypothetical patients [who are] imagining themselves to be in such states.'[14] Indeed, healthy people imagine that eighty-three states of illness would be 'worse than death', and yet, people who are actually in those states rarely take their own lives.[15] If negative events don't hit us as hard as we expect them to, then why do we expect them to? If heartbreaks and calamities can be blessings in disguise, then why are their disguises so convincing? The answer is that the human mind tends to *exploit ambiguity* – and if that phrase seems ambiguous to you, then just keep reading and let me exploit it.

Stop Annoying People

The only thing more difficult than finding a needle in a haystack is finding a needle in a needlestack. When an object is surrounded by similar objects it naturally blends in, and when it is surrounded by dissimilar objects it naturally stands out. Look at figure 16. If you had a stopwatch that counted milliseconds, you'd find that you can locate the letter O in the array on the top (where it is surrounded by numbers) a bit more quickly than you can locate it in the array on the bottom (where it is surrounded by other letters). And that makes sense, because it is harder to find a letter among letters than a letter among numbers. And yet, had I asked you to look for 'zero' instead of 'the letter O', you would have been a bit faster to find it in the array at the bottom than in the array at the top.[16] Now, most of us think that a basic sensory ability such as vision is pretty well explained by its wiring, and if you wanted to understand this ability, you would do well to learn about luminance, contrast, rods, cones, optic nerves, retinas and the like. But once you knew everything there was to know about the physical properties of the arrays shown in figure 16 and everything there was to know about the anatomy of the human eye, you would still not be able to explain why a person can find the circle more quickly in one case than in the other unless you also knew what that person thought the circle *meant*.

Meanings matter for even the most basic psychological processes, and while this may seem perfectly obvious to reasonable people like you and me, ignorance of this perfectly obvious fact sent psychologists on a wild-goose chase that lasted nearly thirty years and produced relatively few geese. For much of the last half of the twentieth century, experimental psychologists timed rats as they ran mazes and observed pigeons as they pecked keys because they believed that the best way to understand behavior was to map the relation between a stimulus and an organism's response to that stimulus. By carefully measuring what an organism did when it was pre-

1	5	9	3	1	5	4	4	2	9
6	8	4	2	1	6	2	2	3	3
9	2	7	6	9	7	5	5	1	1
5	3	7	2	7	6	2	7	8	9
3	7	5	9	6	8	8	2	9	8
4	8	3	1	2	1	6	8	1	8
4	3	4	2	3	9	1	7	0	9
6	2	4	1	8	6	7	5	2	3
7	6	4	2	9	6	5	4	4	5
9	5	2	3	6	7	8	4	5	3

L	G	V	C	L	G	E	E	P	V
I	T	E	P	L	I	P	P	C	C
V	Q	R	I	V	R	G	G	L	L
G	C	R	P	R	I	P	R	T	V
C	R	G	V	I	T	T	P	V	T
E	T	C	L	P	L	I	T	L	T
E	C	E	P	C	V	L	R	O	V
I	P	E	L	T	I	R	G	P	C
R	I	E	P	V	I	G	E	E	G
V	G	Q	C	I	R	T	E	G	C

Fig. 16.

sented with a physical stimulus, such as a light, a sound or a piece of food, psychologists hoped to develop a science that linked observable stimuli to observable behavior without using vague and squishy concepts such as *meaning* to connect them. Alas, this simpleminded project was doomed from the start, because while rats and pigeons may respond to stimuli as they are *presented* in the world, people respond to stimuli as they are *represented* in the mind. Objective stimuli in the world create subjective stimuli in the mind, and it is these subjective stimuli to which people react. For instance, the middle letters in the two words in figure 17 are physically identical stimuli (I promise – I cut and pasted them myself), and yet, most English speakers respond to them differently – see them differently, pronounce them differently, remember them differently – because one represents the letter *H* and the other represents the letter *A*. Indeed, it would be more appropriate to say that one *is* the letter *H* and the other *is* the letter *A* because the identity of an inky squiggle has less to do with how it is objectively constructed and more to do with *how we subjectively interpret it*. Two vertical lines with a crossbar *mean* one thing when flanked by *T* and *E* and they *mean* another thing when flanked by *C* and *T*, and one of the many things that distinguishes us from rats and pigeons is that we respond to the *meanings* of such stimuli and not to the stimuli themselves. That's why my father can get away with calling me "doodlebug" and you can't.

TⱯE
CⱯT

Fig. 17. The middle shape has different meanings in different contexts.

Disambiguating Objects

Most stimuli are ambiguous – that is, they can mean more than one thing – and the interesting question is how we *disambiguate* them – that is, how we know which of a stimulus's many meanings to infer on a particular occasion. Research shows that *context*, *frequency* and *recency* are especially important in this regard.

- Consider *context*. The word *bank* has two meanings in English: 'a place where money is kept' and 'the land on either side of a river'. Yet we never misunderstand sentences such as 'The boat ran into the bank' or 'The robber ran into the bank' because the words *boat* and *robber* provide a context that tells us which of the two meanings of *bank* we should infer in each case.

- Consider *frequency*. Our past encounters with a stimulus provide information about which of its meanings we should embrace. For example, a loan officer is likely to interpret the sentence 'Don't run into the bank' as a warning about how to ambulate through his place of business and not as sound advice about the steering of boats because in the course of a typical day the loan officer hears the word *bank* used more frequently in its financial than in its maritime sense.

- Consider *recency*. Even a sailor is likely to interpret the sentence 'Don't run into the bank' as a reference to a financial institution rather than a river's edge if she recently saw an ad for safe-deposit boxes and thus has the financial meaning of *bank* still active in her mind. Indeed, because I've been talking about banks in this paragraph, I am willing to bet that the spokensentence 'He put a check in the box' causes you to generate a mental image of someone placing a piece of paper in a receptacle and not a mental image of someone making a mark on a questionnaire. (I'm also willing to guess that your interpretation of the title of this section depends on whether you annoyed someone more or less recently than someone annoyed you.)

Unlike rats and pigeons, then, we respond to meanings – and context, frequency and recency are three of the factors that determine which meaning we will infer when we encounter an ambiguous stimulus. But there is another factor of equal importance and greater interest. Like rats and pigeons, each of us has desires, wishes and needs. We are not merely spectators of the world but investors in it, and we often *prefer* that an ambiguous stimulus mean one thing rather than another. Consider, for example, the drawing of a box in figure 18. This object (called the Necker cube after the Swiss crystallographer who discovered it in 1832) is inherently ambiguous, and you can prove this to yourself simply by staring at it for a few seconds. At first, the box appears to be sitting on its side and you have the sense that you're looking out at a box that is *across* from you. The dot is inside the box, at the place where the back panel and the bottom panel meet. But if you stare long enough, the drawing suddenly shifts, the box appears to be standing on its end, and you have the sense that you're looking down on a box that is *below* you. The dot is now perched on the upper right corner of the box. Because this drawing has two equally meaningful interpretations, your brain merrily switches back and forth between them, keeping you mildly entertained until you eventually get dizzy and fall down. But what if one of these meanings were better than the other? That is, what if you *preferred* one of the interpretations of this object? Experiments show that when subjects are rewarded for seeing the box across from them or below them, the orientation for which they were rewarded starts 'popping out' more often and their brains 'hold on' to that interpretation without switching.[17] In other words, when your brain is at liberty to interpret a stimulus in more than one way, it tends to interpret it the way it *wants* to, which is to say that your preferences influence your interpretations of stimuli in just the same way that context, frequency and recency do.

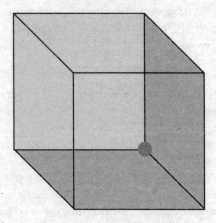

Fig. 18. If you stare at a Necker cube, it will appear to shift its orientation.

This phenomenon is not limited to the interpretation of weird drawings. For example, why is it that you think of yourself as a talented person? (Come on, admit it. You know you do.) To answer this question, researchers asked some volunteers (definers) to write down their definition of *talented* and then to estimate their talent using that definition as a guide.[18] Next, some other volunteers (non-definers) were given the definitions that the first group had written down and were asked to estimate their own talent using those definitions as a guide. Interestingly, the definers rated themselves as more talented than the nondefiners did. Because definers were given the liberty to define the word *talented* any way they wished, they defined it *exactly* the way they wished – namely, in terms of some activity at which they just so happened to excel ('I think *talent* usually refers to *exceptional artistic achievement* like, for example, this painting I just finished', or '*Talent* means *an ability you're born with,* such as being much stronger than other people. Shall I put you down now?'). Definers were able to set the standards for talent, and not coincidentally, they were more likely to meet the standards they set. One of the reasons why most of us think of ourselves as talented, friendly, wise and fair-minded is that these words are the lexical equivalents of a Necker cube, and the human mind naturally exploits each word's ambiguity for its own gratification.

Disambiguating Experience

Of course, the richest sources of exploitable ambiguity are not words, sentences or shapes but the intricate, variegated, multi-dimensional *experiences* of which every human life is a collage. If a Necker cube has two possible interpretations and *talent* has fourteen possible interpretations, then *leaving home* or *falling ill* or *getting a job with the U.S. Postal Service* has hundreds or thousands of possible interpretations. The things that *happen* to us – getting married, raising a child, finding a job, resigning from Congress, going to prison, becoming paralysed – are much more complex than an inky squiggle or a coloured cube, and that complexity creates loads of ambiguity that just begs to be exploited. It doesn't have to beg hard. For example, volunteers in one study were told that they would be eating a delicious but unhealthy ice cream sundae (ice cream eaters), and others were told that they would be eating a bitter but healthy plate of fresh kale (kale eaters).[19] Before actually eating these foods, the researchers asked the volunteers to rate the similarity of a number of foods, including ice cream sundaes, kale and Spam (which everyone considered both unpalatable and unhealthy). The results showed that ice cream eaters thought that Spam was more like kale than it was like ice cream. Why? Because for some odd reason, ice cream eaters were thinking about food in terms of its *taste* – and unlike kale and Spam, ice cream tastes delicious. On the other hand, kale eaters thought that Spam was more like ice cream than it was like kale. Why? Because for some odd reason, kale eaters were thinking about food in terms of its *healthiness* – and unlike kale, ice cream and Spam are unhealthy. The odd reason isn't really so odd. Just as a Necker cube is both across from you and below you, ice cream is both fattening and tasty, and kale is both healthy and bitter. Your brain and my brain easily jump back and forth between these different ways of thinking about the foods because we are merely reading about them. But if we were preparing to *eat* one of them, our brains would automatically exploit the ambiguity of that food's identity and allow us to think of it in a way that pleased us (delicious dessert or nutritious veggie) rather than a way that did not (fattening dessert or bitter veggie). As soon as our *potential* experience becomes our *actual* experience – as soon as we have a stake in its goodness – our brains get busy looking for ways to think about the experience that will allow us to appreciate it.

Because experiences are inherently ambiguous, finding a 'positive view' of an experience is often as simple as finding the 'below-you view' of a Necker cube, and research shows that most people do this well and often. Consumers evaluate kitchen appliances more positively after they buy them,[20] job seekers evaluate jobs more positively after they accept them,[21] and high school students evaluate colleges more positively after they get into them.[22] Racetrack gamblers evaluate their horses more positively when they are leaving the betting window than when they are approaching it,[23] and voters evaluate their candidates more positively when they are exiting the voting booth than when they are entering it.[24] A toaster, a firm, a university, a horse and a senator are all just fine and dandy, but when they become *our* toaster, firm, university, horse and senator they are instantly finer and dandier. Studies such as these suggest that people are quite adept at finding a positive way to view things once those things become their own.

Cooking with Facts

In Voltaire's classic novel *Candide*, Dr Pangloss is a teacher of 'meta-physico-theologo-cosmolo-nigology' who believes he lives in the best of all possible worlds.

> 'It is clear,' he said, 'that things cannot be other than the way they are; for as all things have been created for some end, they must necessarily be created for the best end. For instance, noses were made to support spectacles, hence we wear specta-cles. Legs, as anyone can see, were made for breeches, and so we wear breeches. Stones were made to be shaped into castles; thus My Lord has a fine castle because the greatest baron in the province ought to have the finest house. And because pigs were made to be eaten, we eat pork all year round. So those who say that everything is well are speaking foolishly; they should say that everything is best.'[25]

The research I've described so far seems to suggest that human beings are hopelessly Panglossian; there are more ways to think about experience than there are experiences to think about, and human beings are unusually inventive when it comes to finding the best of all possible ways. And yet, if this is true, then why aren't we all walking around with wide eyes and loopy grins, thanking God for the wonder of haemorrhoids and the miracle of in-laws? Because the mind may be gullible, but it ain't no patsy. The world is *this* way, we wish the world were *that* way, and our experience of the world – how we see it, remember it and imagine it – is a mixture of stark reality and comforting illusion. We can't spare either. If we were to experience the world exactly as it is, we'd be too depressed to get out of bed in the morning, but if we were to experience the world exactly as we want it to be, we'd be too deluded to find our slippers. We may see the world through rose-coloured glasses, but rose-coloured glasses are neither opaque nor clear. They can't be opaque because we need to see the world clearly enough to participate in it – to pilot helicopters, harvest corn, feed babies and all the other stuff that smart mammals need to do in order to survive and thrive. But they can't be clear because we need their rosy tint to motivate us to *design* the helicopters ('I'm sure this thing will fly'), *plant* the corn ('This year will be a banner crop') and *tolerate* the babies ('What a bundle of joy!'). We cannot do without reality and we cannot do without illusion. Each serves a purpose, each imposes a limit on the influence of the other, and our experience of the world is the artful compromise that these tough competitors negotiate.[26]

Rather than thinking of people as hopelessly Panglossian, then, we might think of them as having a *psychological immune system* that defends the mind against unhappiness in much the same way that the *physical immune system* defends the body against illness.[27] This metaphor is unusually appropriate. For example, the physical immune system must strike a balance between two competing needs: the need to recognize and destroy foreign invaders such as viruses and bacteria, and the need to recognize and respect the body's own cells. If the physical immune system is hypoactive, it fails to defend the body against micropredators and we are stricken with infections; but if the physical immune system is hyperactive, it mistakenly defends the body against itself and we are stricken with autoimmune disease. A healthy physical immune system must balance its competing needs and find a way to defend us well – but not *too* well.

Analogously, when we face the pain of rejection, loss, misfortune and failure, the *psychological* immune system must not defend us too well ('I'm perfect and everyone is against me') and must not fail to defend us well enough ('I'm a loser and I ought to be dead'). A *healthy* psychological immune system strikes a balance that allows us to feel good enough to cope with our situation but bad enough to do something about it ('Yeah, that was a lousy performance and I feel crummy about it, but I've got enough confidence to give it a second shot'). We need to be defended – not defenceless or defensive – and thus our minds naturally look for the best view of things while simultaneously insisting that those views stick reasonably closely to the facts. That's why people seek opportunities to think about themselves in positive ways but routinely reject opportunities to think about themselves in *unrealistically* positive ways.[28] For example, college students request new dorm assignments when their current roommates do not think well of them, but they also request new dorm assignments when their current roommates think *too well* of them.[29] No one likes to feel that they are being duped, even when the duping is a pleasure. In order to maintain the delicate balance between reality and illusion, we seek positive views of our experience, but we only allow ourselves to embrace those views when they seem *credible*. So what makes a view seem credible?

Finding Facts

Most of us put a lot of stock in what scientists tell us because we know that scientists reach their conclusions by gathering and analysing facts. If someone asked you why you believe that smoking is bad and jogging is good, or that the earth is round and the galaxy is flat, or that cells are small and molecules are smaller, you would point to the facts. You might need to explain that you do not personally *know* the facts to which you are pointing, but that you do know that at some time in the past, a bunch of very earnest people in white lab coats went out and observed the world with stethoscopes, telescopes and microscopes, wrote down what they observed, analysed what they wrote down and then told the rest of us what to believe about nutrition, cosmology and biology. Scientists are credible because they draw conclusions from observations, and ever since the *empiricists* trumped the *dogmatists* and became the kings of ancient Greek medicine, westerners have had a special reverence for conclusions that are based on things they can see. It isn't surprising, then, that we consider our own views credible when they are based on observable facts but not when they are based on wishes, wants and fancies. We might *like* to believe that everyone loves us, that we will live forever and that high-tech stocks are preparing to make a major comeback, and it would be awfully convenient if we could just push a little button at the base of our skulls and instantly believe as we wanted. But that's not how believing works. Over the course of human evolution, the brain and the eye have developed a contractual relationship in which the brain has agreed to believe what the eye sees and not to believe what the eye denies. So if we are to believe something, then it must be supported by – or at least not blatantly contradicted by – the facts.

If views are acceptable only when they are credible, and if they are credible only when they are based on facts, then how do we achieve positive views of ourselves and our experience? How do we manage to think of ourselves as great drivers, talented lovers and brilliant chefs when the facts of our lives include a pathetic parade of dented cars, disappointed partners and deflated soufflés? The answer is simple: *we cook the facts*. There are many different techniques for collecting, interpreting and analysing facts, and different techniques often lead to different conclusions, which is why scientists disagree about the dangers of global warming, the benefits of supply-side economics and the wisdom of low-carbohydrate diets. Good scientists deal with this complication by choosing the techniques they consider most appropriate and then accepting the conclusions that these techniques produce, regardless of what those conclusions might be. But *bad* scientists take advantage of this complication by choosing techniques that are especially likely to produce the conclusions they favour, thus allowing them to reach favoured conclusions by way of supportive facts. Decades of research suggests that when it comes to collecting and analysing facts about ourselves and our experiences, most of us have the equivalent of an advanced degree in Really Bad Science.

Consider, for instance, the problem of sampling. Because scientists cannot observe every bacterium, comet, pigeon or person, they study small samples that are drawn from these populations. A fundamental rule of good science and common sense is that this sample must be drawn from all parts of the population if it is to tell us about that population. There's really no point in conducting an opinion poll if you're only going to call registered Republicans from Orange County, California or the executive membership of Anarchists Against Organizations Including This One. And yet, that's pretty much what we do when seeking facts that bear on our favoured conclusions.[30] For example, when volunteers in one study were told that they'd scored poorly on an intelligence test and were then given an opportunity to peruse newspaper articles about IQ tests, they spent more time reading articles that questioned the validity of such tests than articles that sanctioned them.[31] When volunteers in another study were given a glowing evaluation by a supervisor, they were more interested in reading background information that praised the supervisor's competence and acumen than background information that impeached it.[32] By controlling the sample of information to which they were exposed, these people indirectly controlled the conclusions they would draw.

You've probably done this yourself. For instance, if you've ever purchased a new automobile, you may have noticed that soon after you made the decision to buy the Honda instead of the Toyota, you began lingering over the Honda advertisements in the weekly newsmagazine and skimming quickly past ads for the competition.[33] If a friend had noticed this and asked you about it, you would probably have explained that you were simply more interested in learning about the car you'd chosen than about the car you didn't. But *learning* is an odd choice of words here because that word usually refers to the balanced acquisition of knowledge, and the kind of learning one does by reading only Honda ads is more than a little lopsided. Ads contain facts about the advantages of the products they describe and not about the disadvantages, and thus your quest for new knowledge would have the interesting side benefit of ensuring that you would be marinated in those facts – and *only* those facts – that confirmed the wisdom of your decision.

Not only do we select favourable facts from magazines, we also select them from memory. For example, in one study, some volunteers were shown evidence indicating that extroverts receive higher salaries and more promotions than introverts do (successful-extrovert group) and other volunteers were shown evidence indicating the opposite (successful-introvert group).[34] When the volunteers were asked to recall specific behaviours from their pasts that would help determine whether they were extroverted or introverted, volunteers in the successful-extrovert group tended to recall the time when they'd brazenly walked up to a complete stranger and introduced themselves, whereas volunteers in the successful-introvert group tended to recall the time when they saw someone they liked but had been too shy to say hello.

Of course, other people – and not memories or magazine ads – are the richest sources of information about the wisdom of our decisions, the extent of our abilities and the irresistible effervescence of our bubbly personalities. Our tendency to expose ourselves to information that supports our favoured conclusions is especially powerful when it comes to choosing the company we keep. You've probably noticed that with the exception of almost no one, nobody picks friends and lovers by random sampling. On the contrary, we spend countless hours and countless dollars carefully arranging our lives to ensure that we are surrounded by people who *like* us, and people who *are* like us. It isn't surprising, then, that when we turn to the people we know for advice and opinions, they tend to confirm our favoured conclusions – either because they share them or because they don't want to hurt our feelings by telling us otherwise.[35] Should the people in our lives occasionally fail to tell us what we want to hear, we have some clever ways of helping them.

For example, studies reveal that people have a penchant for asking questions that are subtly engineered to manipulate the answers they receive.[36] A question such as 'Am I the best lover you've ever had?' is dangerous because it has only one answer that can make us truly happy, but a question such as 'What do you like best about my lovemaking?' is brilliant because it has only one answer that can make us truly miserable. Studies show that people intuitively lean toward asking the questions that are most likely to elicit the answers they want to hear. And when they hear those answers, they tend to believe what they've nudged others to say, which is why 'Tell me you love me' remains such a popular request.[37] In short, we derive support for our preferred conclusions by listening to the words that we put in the mouths of people who have already been preselected for their willingness to say what we want to hear.

And it gets worse – because most of us have ways of making other people confirm our favoured conclusions without ever engaging them in conversation. Consider this: to be a great driver, lover or chef, we don't need to be able to parallel park while blindfolded, make ten thousand maidens swoon with a single pucker or create a *pâte feuilletée* so intoxicating that the entire population of France instantly abandons its national cuisine and swears allegiance to our kitchen. Rather, we simply need to park, kiss and bake better than most other people do. How do we know how well most other people do? Why, we look around, of course – but in order to make sure that we see what we want to see, we look around *selectively*.[38] For example, volunteers in one study took a test that ostensibly measured their social sensitivity and were then told that they had flubbed the majority of the questions.[39] When these volunteers were then given an opportunity to look over the test results of other people who had performed better or worse than they had, they ignored the tests of the people who had done better and instead spent their time looking over the tests of the people who had done worse. Getting a C- isn't so bad if one compares oneself exclusively to those who got a D.

This tendency to seek information about those who have done more poorly than we have is especially pronounced when the stakes are high. People with life-threatening illnesses such as cancer are particularly likely to compare themselves with those who are in worse shape,[40] which explains why 96 per cent of the cancer patients in one study claimed to be in better health than the average cancer patient.[41] And if we can't *find* people who are doing more poorly than we are, we may go out and *create* them. Volunteers in one study took a test and were then given the opportunity to provide hints that would either help or hinder a friend's performance on the same test.[42] Although volunteers helped their friends when the test was described as a game, they actively hindered their friends when the test was described as an important measure of intellectual ability. Apparently, when our friends do not have the good taste to come in last so that we can enjoy the good taste of coming in first, we give them a friendly push in the appropriate direction. Once we've successfully sabotaged their performances and ensured their failure, they become the perfect standard for comparison. The bottom line is this: the brain and the eye may have a contractual relationship in which the brain has agreed to believe what the eye sees, but in return the eye has agreed to look for what the brain wants.

Challenging Facts

Whether by choosing information or informants, our ability to cook the facts that we encounter helps us establish views that are both positive and credible. Of course, if you've ever discussed a football game, a political debate or the six o'clock newscast with someone from the other side of the aisle, you've already discovered that even when people *do* encounter facts that disconfirm their favoured conclusions, they have a knack for ignoring them, forgetting them or seeing them differently than the rest of us do. When Dartmouth and Princeton students see the same football game, both sets of students claim that the facts clearly show that the other school's team was responsible for the unsportsmanlike conduct.[43] When Democrats and Republicans see the same presidential debate on television, both sets of viewers claim that the facts clearly show that their candidate was the winner.[44] When pro-Israeli and pro-Arab viewers see identical samples of Middle East news coverage, both proponents claim that the facts clearly show that the press was biased against their side.[45] Alas, the only thing these facts *clearly* show is that people tend to see what they want to see.

Inevitably, however, there will be times when the unkind facts are just too obvious to set aside. When our team's defensive tackle is caught wearing brass knuckles, or when our candidate confesses to embezzlement on national television, we find it difficult to overlook or forget such facts. How do we manage to maintain a favoured conclusion when the brute facts just won't cooperate? Although the word *fact* seems to suggest a sort of unquestionable irrefutability, facts are actually nothing more than conjectures that have met a certain standard of proof. If we set that standard high enough, then nothing can ever be proved, including the 'fact' of our own existence. If we set the standard low enough, then all things are true and equally so. Because nihilism and postmodernism are both such unsatisfying philosophies, we tend to set our standard of proof somewhere in the middle. No one can say precisely where that standard should be set, but one thing we do know is that wherever we set it, we must keep it in the same place when we evaluate the facts we favour and the facts we don't. It would be unfair for teachers to give the students they like easier exams than those they dislike, for federal regulators to require that foreign products pass stricter safety tests than domestic products or for judges to insist that the defence attorney make better arguments than the prosecutor.

And yet, this is just the sort of uneven treatment most of us give to facts that confirm and disconfirm our favoured conclusions. In one study, volunteers were asked to evaluate two pieces of scientific research on the effectiveness of capital punishment as a deterrent.[46] They were shown one research study that used the 'between-states technique' (which involved comparing the crime rates of states that had capital punishment with the crime rates of states that did not) and one research study that used the 'within-states technique' (which involved comparing the crime rates of a single state before and after it instituted or outlawed capital punishment). For half the volunteers, the between-states study concluded that capital punishment was effective and the within-states study concluded it was not. For the other half of the volunteers, these conclusions were reversed. The results showed that volunteers favoured whichever technique produced the conclusion that verified their own personal political ideologies. When the within-states technique produced an unfavourable conclusion, volunteers immediately recognized that within-states comparisons are worthless because factors such as employment and income vary over time, and thus crime rates in one decade (the 1980s) can't be compared with crime rates in another decade (the 1990s). But when the between-states technique produced an unfavourable conclusion, volunteers immediately recognized that between-states comparisons are worthless because factors such as employment and income vary with geography, and thus crime rates in one place can't be compared with crime rates in another place.[47] Clearly, volunteers set the methodological bar higher for studies that disconfirmed their favoured conclusions. This same technique allows us to achieve and maintain a positive and credible view of ourselves and our experiences. For example, volunteers in one study were told that they had performed very well or very poorly on a social-sensitivity test and were then asked to assess two scientific reports – one that suggested the test was valid and one that suggested it was not.[48] Volunteers who had performed well on the test believed that the studies in the validating report used sounder scientific methods than did the studies in the invalidating report, but volunteers who performed poorly on the test believed precisely the opposite.

When facts challenge our favoured conclusion, we scrutinize them more carefully and subject them to more rigorous analysis. We also require a lot more of them. For example, how much information would you require before you were willing to conclude that someone was intelligent? Would their high school transcripts be enough? Would an IQ test suffice? Would you need to know what their teachers and employers thought of them? Volunteers in one study were asked to evaluate the intelligence of another person, and they required considerable evidence before they were willing to conclude that the person was truly smart. But interestingly, they required much *more* evidence when the person was an unbearable pain in the ass than when the person was funny, kind and friendly.[49] When we *want* to believe that someone is smart, then a single letter of recommendation may suffice; but when we *don't want* to believe that person is smart, we may demand a thick manila folder full of transcripts, tests and testimony.

Precisely the same thing happens when we want or don't want to believe something about ourselves. For instance, volunteers in one study were invited to take a medical test that would supposedly tell them whether they did or did not have a dangerous enzyme deficiency that would predispose them to pancreatic disorders.[50] The volunteers placed a drop of their saliva on a strip of ordinary paper that the researchers falsely claimed was a medical test strip. Some volunteers (positive-testers) were told that if the strip turned green in ten to sixty seconds, then they had the enzyme deficiency. Other volunteers (negative-testers) were told that if the strip turned green in ten to sixty seconds, then they *didn't* have the enzyme deficiency. Although the strip was an ordinary piece of paper and hence never turned green, the negative-testers waited much longer than the positive-testers before deciding that the test was complete. In other words, the volunteers gave the test strip plenty of time to prove that they were well but much less time to prove that they were ill. Apparently it doesn't take much to convince us that we are smart and healthy, but it takes a whole lotta facts to convince us of the opposite. We ask whether facts *allow* us to believe our favoured conclusions and whether they *compel* us to believe our disfavoured conclusions.[51] Not surprisingly, disfavoured conclusions have a much tougher time meeting this more rigorous standard of proof.[52]

Onward

In July 2004, the City Council of Monza, Italy, took the unusual step of banning goldfish bowls. They reasoned that goldfish should be kept in rectangular aquariums and not in round bowls because 'a fish kept in a bowl has a distorted view of reality and suffers because of this'.[53] No mention was made of the bland diet, the noisy pump or the silly plastic castles. No, the problem was that round bowls deform the visual experience of their inhabitants, and goldfish have the fundamental right to see the world as it really is. The good counselors of Monza did not suggest that human beings should enjoy the same right, perhaps because they knew that our distorted views of reality are not so easily dispelled, or perhaps because they understood that we suffer less with them than we would without them. Distorted views of reality are made possible by the fact that experiences are ambiguous – that is, they can be credibly viewed in many ways, some of which are more positive than others. To ensure that our views are credible, our brain accepts what our eye sees. To ensure that our views are positive, our eye looks for what our brain wants. The conspiracy between these two servants allows us to live at the fulcrum of stark reality and comforting illusion. So what does all of this have to do with forecasting our emotional futures? As we are about to see, we may live at the fulcrum of reality and illusion, but most of us don't know our own address.

CHAPTER 9

Immune to Reality

Upon my back, to defend my belly; upon my wit, to
defend my wiles; upon my secrecy, to defend mine
honesty; my mask, to defend my beauty.

Shakespeare, *Troilus and Cressida*

ALBERT EINSTEIN MAY HAVE BEEN the greatest genius of the
twentieth century, but few people know that he came *this* close to
losing that distinction to a horse. Wilhelm von Osten was a retired
schoolteacher who in 1891 claimed that his stallion, whom he called
Clever Hans, could answer questions about current events, mathematics and a host of other topics by tapping the ground with his
foreleg. For instance, when Osten would ask Clever Hans to add
three and five, the horse would wait until his master had finished
asking the question, tap eight times, then stop. Sometimes, instead
of *asking* a question, Osten would write it on a card and hold it up
for Clever Hans to read, and the horse seemed to understand written
language every bit as well as it understood speech. Clever Hans
didn't get *every* question right, of course, but he did much better
than anyone else with hooves, and his public performances were so
impressive that he soon became the toast of Berlin. But in 1904 the
director of the Berlin Psychological Institute sent his student, Oskar
Pfungst, to look into the matter more carefully, and Pfungst noticed
that Clever Hans was much more likely to give the wrong answer
when Osten was standing behind the horse than when he was in

front of it, or when Osten himself did not know the answer to the question the horse had been asked. In a series of experiments, Clever Pfungst was able to show that Clever Hans could indeed read – but that what he could read was Osten's body language. When Osten bent slightly, Clever Hans would start tapping, and when Osten straightened up, or tilted his head a bit, or faintly raised an eyebrow, Clever Hans would stop. In other words, Osten was signaling Clever Hans to start and stop tapping at just the right moments to create the illusion of horse sense.

Clever Hans was no genius, but Osten was no fraud. Indeed, he'd spent years patiently talking to his horse about mathematics and world affairs, and he was genuinely shocked and dismayed to learn that he had been fooling himself, as well as everyone else. The deception was elaborate and effective, but it was perpetrated unconsciously, and in this Osten was not unique. When we expose ourselves to favourable facts, notice and remember favourable facts, and hold favourable facts to a fairly low standard of proof, we are generally no more aware of our subterfuge than Osten was of his. We may refer to the processes by which the psychological immune system does its job as 'tactics' or 'strategies', but these terms – with their inevitable connotations of planning and deliberation – should not cause us to think of people as manipulative schemers who are consciously *trying* to generate positive views of their own experience. On the contrary, research suggests that people are *typically unaware* of the reasons why they are doing what they are doing,[1] but when asked for a reason, they readily supply one.[2] For example, when volunteers watch a computer screen on which words appear for just a few milliseconds, they are unaware of seeing the words and are unable to guess which words they saw. But they are influenced by them. When the word *hostile* is flashed, volunteers judge others negatively.[3] When the word *elderly* is flashed, volunteers walk slowly.[4] When the word *stupid* is flashed, volunteers perform poorly on tests.[5] When these volunteers are later asked to explain *why* they judged, walked or scored the way they did, two things happen: First, they don't know, and second, they do not say, 'I don't know.' Instead, their brains quickly consider the facts of which they *are* aware ('I walked slowly') and draw the same kinds of plausible but mistaken inferences about themselves that an observer would probably draw about them ('I'm tired').[6]

When we cook facts, we are similarly unaware of why we are doing it, and this turns out to be a good thing, because *deliberate* attempts to generate positive views ("There must be *something* good about bankruptcy, and I'm not leaving this chair until I discover it") contain the seeds of their own destruction. Volunteers in one study listened to Stravinsky's *Rite of Spring*.[7] Some were told to listen to the music, and others were told to listen to the music while consciously trying to be happy. At the end of the interlude, the volunteers who had tried to be happy were in a *worse* mood than were the volunteers who had simply listened to the music. Why? Two reasons. First, we may be able deliberately to generate positive views of our own experiences if we close our eyes, sit very still and do nothing else,[8] but research suggests that if we become even slightly distracted, these deliberate attempts tend to backfire and we end up feeling worse than we did before.[9] Second, deliberate attempts to cook the facts are so transparent that they make us feel cheap. Sure, we *want* to believe that we're better off without the fiancée who left us standing at the altar, and we *will* feel better soon as we begin to discover facts that support this conclusion ('She was never really right for me, was she, Mum?'), but the process by which we discover those facts must *feel* like a discovery and not like a snow job. If we *see* ourselves cooking the facts ('If I phrase the question just this way and ask nobody but Mum, I stand a pretty good chance of having my favoured conclusion confirmed'), then the jig is up and *self-deluded* joins *jilted* in our list of pitiful qualities. For positive views to be credible, they must be based on facts that we *believe* we have come upon honestly. We accomplish this by unconsciously cooking the facts and then consciously consuming them. The diner is in the dining room, but the chef is in the basement. The benefit of all this unconscious cookery is that it works; but the cost is that it makes us strangers to ourselves. Let me show you how.

Looking Forward to Looking Backward

To my knowledge, no one has ever done a systematic study of people who've been left standing at the altar by a cold-footed fiancé. But I'm willing to bet a good bottle of wine that if you rounded up a healthy sample of almost-brides and nearly grooms and asked them whether they would describe the incident as 'the worst thing that ever happened to me' or 'the best thing that ever happened to me', more would endorse the latter description than the former. And I'll bet an entire *case* of that wine that if you found a sample of people who'd never been through this experience and asked them to predict which of all their possible future experiences they are most likely to look back on as 'the best thing that ever happened to me', not one of them will list 'getting jilted'. Like so many things, getting jilted is more painful in prospect and more rosy in retrospect. When we contemplate being hung out to dry this way, we naturally generate the most dreadful possible view of the experience; but once we've actually *been* heartbroken and humiliated in front of our family, friends and florists, our brains begin shopping for a less dreadful view – and as we've seen, the human brain is one smart shopper. However, because our brains do their shopping unconsciously, we tend not to realize they will do it at all, hence we blithely assume that the dreadful view we have when we look forward to the event is the dreadful view we'll have when we look back on it. In short, we do not realize that our views will change because we are normally unaware of the processes that change them.

This fact can make it quite difficult to predict one's emotional future. In one study, volunteers were given the opportunity to apply for a good-paying job that involved nothing more than tasting ice cream and making up funny names for it.[10] The application procedure required the volunteer to undergo an on-camera interview. Some of the volunteers were told that their interview would be seen by a judge who had sole discretionary authority to decide whether they would be hired (judge group). Other volunteers were told that their interview would be seen by a jury whose members would vote to decide whether the volunteer should be hired (jury group). Volunteers in the jury group were told that as long as *one* juror voted for them, they would get the job – and thus the only circumstance under which they would *not* get the job was if the jury voted unanimously against them. All of the volunteers then underwent an interview, and all predicted how they would feel if they didn't get the job. A few minutes later, the researcher came into the room and explained apologetically that after careful deliberation, the judge or jury had decided that the volunteer just wasn't quite right for the job. The researcher then asked the volunteers to report how they felt.

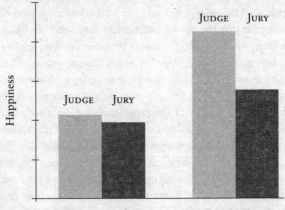

Fig. 19. Volunteers were happier when they were rejected by a capricious judge than by a unanimous jury *(bars on right)*. But they could not foresee this moments before it happened *(bars on left)*.

The results of the study are shown in figure 19. As the bars on the left show, volunteers in the two groups expected to feel equally unhappy. After all, rejection is a major whack on the nose, and we expect it to hurt whether the whacker is a judge, a jury or a gang of Orthodox rabbis. And yet, as the bars on the right show, the whacks hurt more when they were administered by a jury than by a judge. Why? Well, just imagine that you've applied for a job as a swimsuit model, which requires that you don something skimpy and parade back and forth in front of some gimlet-eyed twit in a three-dollar suit. If the twit looked you over, shook his head, and said, 'Sorry, but you're not model material', you'd probably feel bad. For a minute or two. But this is the sort of interpersonal rejection that everyone experiences from time to time, and after a few minutes, most of us get over it and on with our lives. We do this quickly because our psychological immune systems have no trouble finding ways to exploit the ambiguity of this experience and soften its sting: 'The guy wasn't paying attention to my extraordinary pivot' or 'He's one of those weirdos who prefers height to weight' or 'I'm supposed to take fashion advice from a guy with a suit like *that*?'

But now imagine that you've just modelled the skimpy thing for a whole roomful of people – some men, some women, some old, some young – and they all look you over and shake their heads in unison. You'd probably feel bad. Truly bad. Humiliated, hurt and confused. You'd probably hurry offstage with a warm feeling in your ears, a tight feeling in your throat and a wet feeling in your eyes. Being rejected by a large and diverse group of people is a demoralizing experience because it is so thoroughly unambiguous, and hence it is difficult for the psychological immune system to find a way to think about it that is both positive and credible. It's easy to blame failure on the eccentricities of a judge, but it's much more difficult to blame failure on the eccentricities of a unanimous jury. Claims such as 'a synchronized mass blink caused ninety-four people to miss my pivot at precisely the same moment' are just not credible. Similarly, volunteers in this study found it easier to blame their rejection on an idiosyncratic judge than on a panel of jurors, which is why they felt worse when they were rejected by a jury.

Now, all this may seem painfully obvious to you as you contemplate the results of this study from the comfort of your sofa, but allow me to suggest that it is painfully obvious only after someone has taken pains to point it out to you. Indeed, if it were really painfully obvious, then why were a bunch of smart volunteers *unable to predict that it would happen just a few minutes before it did?* Why didn't the volunteers realize that they would have more success blaming a judge than a jury? Because when volunteers were asked to predict their emotional reactions to rejection, they imagined its sharp sting. Period. They did not go on to imagine how their brains might try to relieve that sting. Because they were unaware that they would alleviate their suffering by blaming those who caused it, it never occurred to them that they would be more successful if a single person were to blame rather than an entire group. Other studies have confirmed this general finding. For example, people *expect* to feel equally bad when a tragic accident is the result of human negligence as when it is the result of dumb luck, but they *actually* feel worse when luck is dumb and no one is blameworthy.[11]

Ignorance of our psychological immune systems causes us to mispredict the circumstances under which we will blame others, but it also causes us to mispredict the circumstances under which we will blame ourselves.[12] Who can forget the scene at the end of the 1942 film *Casablanca* in which Humphrey Bogart and Ingrid Bergman are standing on the tarmac as she tries to decide whether to stay in Casablanca with the man she loves or board the plane and leave with her husband? Bogey turns to Bergman and says: 'Inside we both know you belong with Victor. You're part of his work, the thing that keeps him going. If that plane leaves the ground and you're not with him, you'll regret it. Maybe not today. Maybe not tomorrow. But soon and for the rest of your life.'[13]

This thin slice of melodrama is among the most memorable scenes in the history of cinema – not because it is particularly well acted or particularly well written but because most of us have stood on that same runway from time to time. Our most consequential choices – whether to marry, have children, buy a house, enter a profession, move abroad – are often shaped by how we imagine our future regrets ('Oh no, I forgot to have a baby!'). Regret is an emotion we feel when we blame ourselves for unfortunate outcomes that might have been prevented had we only behaved differently in the past, and because that emotion is decidedly unpleasant, our behaviour in the present is often designed to preclude it.[14] Indeed, most of us have elaborate theories about when and why people feel regret, and these theories allow us to avoid the experience. For instance, we expect to feel more regret when we learn about alternatives to our choices than when we don't,[15] when we accept bad advice than when we reject good advice,[16] when our bad choices are unusual rather than conventional,[17] and when we fail by a narrow margin rather than by a wide margin.[18]

But sometimes these theories are wrong. Consider this scenario. You own shares in Company A. During the past year you considered switching to stock in Company B but decided against it. You now find that you would have been better off by $1,200 if you had switched to the stock of Company B. You also owned shares in Company C. During the past year you switched to stock in Company D. You now find out that you'd have been better off by $1,200 if you kept your stock in Company C. Which error causes you more regret? Studies show that about nine out of ten people expect to feel more regret when they foolishly switch stocks than when they foolishly fail to switch stocks, because most people think they will regret foolish actions more than foolish inactions.[19] But studies also show that nine out of ten people are wrong. Indeed, in the long run, people of every age and in every walk of life seem to regret *not* having done things much more than they regret things they *did*, which is why the most popular regrets include not going to college, not grasping profitable business opportunities and not spending enough time with family and friends.[20]

But why do people regret inactions more than actions? One reason is that the psychological immune system has a more difficult time manufacturing positive and credible views of inactions than of actions.[21] When our action causes us to accept a marriage proposal from someone who later becomes an axe murderer, we can console ourselves by thinking of all the things we learned from the experience ('Collecting hatchets is not a healthy hobby'). But when our inaction causes us to reject a marriage proposal from someone who later becomes a movie star, we can't console ourselves by thinking of all the things we learned from the experience because . . . well, there wasn't one. The irony is all too clear: because we do not realize that our psychological immune systems can rationalize an excess of courage more easily than an excess of cowardice, we hedge our bets when we should blunder forward. As students of the silver screen recall, Bogart's admonition about future regret led Bergman to board the plane and fly away with her husband. Had she stayed with Bogey in Casablanca, she would probably have felt just fine. Not right away, perhaps, but soon, and for the rest of her life.

Little Triggers

Civilized people have learned the hard way that a handful of iniquitous individuals can often cause more death and destruction than an invading army. If an enemy were to launch hundreds of airplanes and missiles against the United States, the odds are that none would reach its target because an offensive strike of that magnitude would trigger America's defensive systems, which are presumably adequate to quash the threat. On the other hand, were an enemy to launch seven guys with baggy pants and baseball caps, those men might well reach their targets and detonate bombs, release toxins or fly hijacked airplanes into tall buildings. Terrorism is a strategy based on the idea that the best offence is the one that fails to trigger the best defence, and small-scale incursions are less likely to set off the alarm bells than are large-scale assaults. Although it is possible to design a defensive system that counters even the smallest threat (e.g., electrified borders, a travel ban, electronic surveillance, random searches), such systems are extraordinarily costly, in terms of both the resources required to run them and the number of false alarms they produce. A system like that would be an exercise in overkill. To be effective, a defensive system must respond to threats; but to be practical, it must respond only to threats that exceed some *critical threshold* – which means that threats that fall short of the critical threshold may have a destructive potential that belies their diminutive size. Unlike large threats, small threats can sneak in under the radar.

The Intensity Trigger

The psychological immune system is a defensive system, and it obeys this same principle. When experiences make us feel sufficiently unhappy, the psychological immune system cooks facts and shifts blame in order to offer us a more positive view. But it doesn't do this *every* time we feel the slightest tingle of sadness, jealousy, anger or frustration. Failed marriages and lost jobs are the kinds of large-scale assaults on our happiness that trigger our psychological defences, but these defences are not triggered by broken pencils, stubbed toes or slow elevators. Broken pencils may be annoying, but they do not pose a grave threat to our psychological well-being and hence do not trigger our psychological defences. The paradoxical consequence of this fact is that it is sometimes more difficult to achieve a positive view of a *bad* experience than of a *very bad* experience.

For example, volunteers in one study were students who were invited to join an extracurricular club whose initiation ritual required that they receive three electric shocks.[22] Some of the volunteers had a truly dreadful experience because the shocks they received were quite severe (severe-initiation group), and others had a slightly unpleasant experience because the shocks they received were relatively mild (mild-initiation group). Although you might expect people to dislike anything associated with physical pain, the volunteers in the severe-initiation group actually liked the club more. Because these volunteers suffered greatly, the intensity of their suffering triggered their defensive systems, which immediately began working to help them achieve a credible and positive view of their experience. It isn't easy to find such a view, but it can be done. For example, physical suffering is bad ('Oh my God, that *really* hurt!'), but it isn't *entirely* bad if the thing one suffers for is extremely valuable ('But I'm joining a *very* elite group of *very* special people'). Indeed, research shows that when people are given electric shocks, they actually feel *less pain* when they believe they are suffering for something of great value.[23] The intense shocks were unpleasant enough to trigger the volunteers' psychological defences, but the mild shocks were not, hence the volunteers valued the club most when its initiation was most painful.[24] If you've managed to forgive your spouse for some egregious transgression but still find yourself miffed about the dent in the garage door or the trail of dirty socks on the staircase, then you have experienced this paradox.

Intense suffering triggers the very processes that eradicate it, while mild suffering does not, and this counterintuitive fact can make it difficult for us to predict our emotional futures. For example, would it be worse if your best friend insulted you or insulted your cousin? As much as you may like your cousin, it's a pretty good bet that you like yourself more, hence you probably think that it would be worse if the epithet were hurled your way. And you're right. It *would* be worse. At first. But if intense suffering triggers the psychological immune system and mild suffering does not, then over time you should be more likely to generate a positive view of an insult that was directed at you ('Felicia called me a pea-brain . . . boy, she can really crack me up sometimes') than one that was directed at your cousin ('Felicia called Cousin Dwayne a pea-brain . . . I mean, she's *right*, of course, but it wasn't very nice of her to say'). The irony is that you may ultimately feel better when you are the *victim* of an insult than when you are a *bystander* to it.

This possibility was tested in a study in which two volunteers took a personality test and then *one* of them received feedback from a psychologist.[25] The feedback was professional, detailed and unrelentingly negative. For example, it contained statements such as 'You have few qualities that distinguish you from others' and 'People like you primarily because you don't threaten their competence.' Both of the volunteers read the feedback and then reported how much they liked the psychologist who had written it. Ironically, the volunteer who was the *victim* of the negative feedback liked the psychologist *more* than did the volunteer who was merely a *bystander* to it. Why? Because bystanders were miffed ('Man, that was a really crummy thing to do to the other volunteer'), but they were not devastated, hence their psychological immune systems did nothing to ameliorate their mildly negative feelings. But victims *were* devastated ('Yikes, I'm a certified loser!'), hence their brains quickly went shopping for a positive view of the experience ('But now that I think of it, that test could only provide a small glimpse into my very complex personality, so I rather doubt it means much'). Now here's the important finding: when a new group of volunteers was asked to *predict* how much they would like the psychologist, they predicted that they would like the psychologist *less* if they were victims than if they were bystanders. Apparently, people are not aware of the fact that their defences are more likely to be triggered by intense than mild suffering, thus they mispredict their own emotional reactions to misfortunes of different sizes.

The Inescapability Trigger

Intense suffering is one factor that can trigger our defences and thus influence our experiences in ways we don't anticipate. But there are others. For example, why do we forgive our siblings for behavior we would never tolerate in a friend? Why aren't we disturbed when the president does something that would have kept us from voting for him had he done it before the election? Why do we overlook an employee's chronic tardiness but refuse to hire a job seeker who is two minutes late for the interview? One possibility is that blood is thicker than water, flags were made to be rallied around and first impressions matter most. But another possibility is that we are more likely to look for and find a positive view of the things we're *stuck with* than of the things we're not.[26] Friends come and go, and changing candidates is as easy as changing socks. But siblings and presidents are *ours*, for better or for worse, and there's not much we can do about it once they've been born or elected. When the experience we are having is not the experience we *want* to be having, our first reaction is to go out and have a different one, which is why we return unsatisfactory rental cars, check out of bad hotels and stop hanging around with people who pick their noses in public. It is only when we cannot *change the experience* that we look for ways to *change our view of the experience*, which is why we love the clunker in the driveway, the shabby cabin that's been in the family for years and Uncle Sheldon despite his predilection for nasal spelunking. We find silver linings only when we must, which is why people experience an increase in happiness when genetic tests reveal that they *don't* have a dangerous genetic defect, or when the tests reveal that they *do* have a dangerous genetic defect, but *not* when the tests are inconclusive.[27] We just can't make the best of a fate until it is inescapably, inevitably and irrevocably ours.

Inescapable, inevitable and irrevocable circumstances trigger the psychological immune system, but, as with the intensity of suffering, people do not always recognize that this will happen. For example, college students in one study signed up for a course in black-and-white photography.[28] Each student took a dozen photographs of people and places that were personally meaningful, then reported for a private lesson. In these lessons, the teacher spent an hour or two showing students how to print their two best photographs. When the prints were dry and ready, the teacher said that the student could keep one of the photographs but that the other would be kept on file as an example of student work. Some students (inescapable group) were told that once they had chosen a photograph to take home, they would not be allowed to change their minds. Other students (escapable group) were told that once they had chosen a photograph to take home, they would have several days to change their minds – and if they did, the teacher would gladly swap the photograph they'd taken home for the one they'd left behind. Students made their choices and took one of their photographs home. Several days later, the students responded to a survey asking them (among other things) how much they liked their photographs. The results showed that students in the escapable group liked their photograph *less* than did students in the inescapable group. Interestingly, when a new group of students was asked to *predict* how much they would like their photographs if they were or were not given the opportunity to change their minds, these students predicted that escapability would have no influence whatsoever on their satisfaction with the photograph. Apparently, inescapable circumstances trigger the psychological defences that enable us to achieve positive views of those circumstances, but we do not anticipate that this will happen.

Our failure to anticipate that inescapability will trigger our psychological immune systems (hence promote our happiness and satisfaction) can cause us to make some painful mistakes. For example, when a new group of photography students was asked whether they would prefer to have or not to have the opportunity to change their minds about which photograph to keep, the vast majority preferred to have that opportunity – that is, the vast majority of students preferred to enrol in a photography course in which they would ultimately be dissatisfied with the photograph they produced. Why would anyone prefer less satisfaction to more? No one does, of course, but most people do seem to prefer more freedom to less. Indeed, when our freedom to make up our minds – or to change our minds once we've made them up – is threatened, we experience a strong impulse to reassert it,[29] which is why retailers sometimes threaten your freedom to own their products with claims such as 'Limited stock' or 'You must order by midnight tonight.'[30] Our fetish for freedom leads us to patronize expensive department stores that allow us to return merchandise rather than attend auctions that don't, to lease cars at a dramatic markup rather than buying them at a bargain, and so on.

Most of us will pay a premium today for the opportunity to change our minds tomorrow, and sometimes it makes sense to do so. A few days spent test-driving a little red roadster tells us a lot about what it might be like to own one, thus it is sometimes wise to pay a modest premium for a contract that includes a short refund period. But if keeping our options open has benefits, it also has costs. Little red roadsters are naturally cramped, and while the committed owner will find positive ways to view that fact ('Wow! It feels like a fighter jet!'), the buyer whose contract includes an escape clause may not ('This car is so tiny. Maybe I should return it'). Committed owners attend to a car's virtues and overlook its flaws, thus cooking the facts to produce a banquet of satisfaction, but the buyer for whom escape is still possible (and whose defences have not yet been triggered) is likely to evaluate the new car more critically, paying special attention to its imperfections as she tries to decide whether to keep it. The costs and benefits of freedom are clear – but alas, they are not equally clear: we have no trouble anticipating the advantages that freedom may provide, but we seem blind to the joys it can undermine.[31]

Explaining Away

If you've ever puked your guts out shortly after eating chilli con carne and found yourself unable to eat it again for years, you have a pretty good idea of what it's like to be a fruit fly. No, fruit flies don't eat chilli, and no, fruit flies don't puke. But they do associate their best and worst experiences with the circumstances that accompanied and preceded them, which allows them to seek or avoid those circumstances in the future. Expose a fruit fly to the odor of tennis shoes, give it a very tiny electric shock, and for the rest of its very tiny life it will avoid places that smell tennis-shoey. The ability to associate pleasure or pain with its circumstances is so vitally important that nature has installed that ability in every one of her creatures, from *Drosophila melanogaster* to Ivan Pavlov.

But if that ability is necessary for creatures like us, it certainly isn't sufficient, because the kind of learning it enables is far too limited. If an organism can do no more than associate particular experiences with particular circumstances, then it can learn only a very small lesson, namely, to seek or avoid those particular circumstances in the future. A well-timed shock may teach a fruit fly to avoid the tennis-shoe smell, but it won't teach it to avoid the smell of snowshoes, ballet slippers, Manolo Blahniks or a scientist armed with a miniature stun gun. To maximize our pleasures and minimize our pains, we must be able to associate our experiences with the circumstances that produced them, but we must also be able to *explain* how and why those circumstances produced the experiences they did. If we feel nauseous after a few turns on the Ferris wheel and our explanation involves poor equilibrium, then we avoid Ferris wheels in the future – just as a fruit fly would. But unlike a fruit fly, we also avoid some things that are *not* associated with our nauseating experience (such as bungee jumping and sailboats) and we do *not* avoid some things that *are* associated with our nauseating experience (such as hurdy-gurdy music and clowns). Unlike a mere association, an explanation allows us to identify particular aspects of a circumstance (spinning) as the *cause* of our experience, and other aspects (music) as irrelevant. In so doing, we learn more from our upchucks than a fruit fly ever could.

Explanations allow us to make full use of our experiences, but they also change the nature of those experiences. As we have seen, when experiences are unpleasant, we quickly move to explain them in ways that make us feel better ("I didn't get the job because the judge was biased against people who barf on Ferris wheels"). And indeed, studies show that the mere act of explaining an unpleasant event can help to defang it. For example, simply writing about a trauma – such as the death of a loved one or a physical assault – can lead to surprising improvements in both subjective well-being and

physical health (e.g., fewer visits to the physician and improved production of viral antibodies).[32] What's more, the people who experience the greatest benefit from these writing exercises are those
whose writing contains an *explanation* of the trauma.[33]

But just as explanations ameliorate the impact of *unpleasant*
events, so too do they ameliorate the impact of *pleasant* events. For
example, college students volunteered for a study in which they
believed they were interacting in an online chat room with students
from other universities.[34] In fact, they were actually interacting with
a sophisticated computer program that simulated the presence of
other students. After the simulated students had provided the real
student with information about themselves ('Hi, I'm Eva, and I like
to do volunteer work'), the researcher pretended to ask the simulated students to decide which of the people in the chat room they
liked most, to write a paragraph explaining why, and then to send it
to that person. In just a few minutes, something remarkable happened: the real student received e-mail messages from *every one* of
the simulated students indicating that they liked the real student
best! For example, one simulated message read: 'I just felt that something clicked between us when I read your answers. It's too bad
we're not at the same school!' Another read: 'You stood out as the
one I would like the most. I was especially interested in the way you
described your interests and values.' A third read: 'I wish I could talk
with you directly because . . . I'd ask you if you like being around
water (I love water-skiing) and if you like Italian food (it's my
favourite).'

Now, here's the catch: some real students (informed group) received e-mail that allowed them to know *which* simulated student
wrote each of the messages, and other real students (uninformed
group) received e-mail messages that had been stripped of that
identifying information. In other words, every real student received
exactly the same e-mail messages indicating that they had won
the hearts and minds of all the simulated people in the chat room,
but only real students in the informed group knew *which* simulated
individual had written each of the messages. Hence, real students
in the informed group were able to generate explanations for their

good fortune ('Eva appreciates my values because we're both involved with Habitat for Humanity, and it makes sense that Catarina would mention Italian food'), whereas real students in the uninformed group were not ('Someone appreciates my values . . . I wonder who? And why would anyone mention Italian food?'). The researchers measured how happy the real students were immediately after receiving these messages and then again fifteen minutes later. Although real students in both groups were initially delighted to have been chosen as everyone's best friend, only the real students in the uninformed group remained delighted fifteen minutes later. If you've ever had a secret admirer, then you understand why real students in the uninformed group remained on cloud nine while real students in the informed group quickly descended to clouds two through five.

Unexplained events have two qualities that amplify and extend their emotional impact. First, they strike us as rare and unusual.[35] If I told you that my brother, my sister and I were all born on the same day, you'd probably consider that a rare and unusual occurrence. Once I explained that we were triplets, you'd find it considerably less so. In fact, just about *any* explanation I offered ('By *same day* I meant we were all born on a Thursday' or 'We were all delivered by cesarean section, so Mum and Dad timed our births for maximum tax benefits') would tend to reduce the amazingness of the coincidence and make the event seem more probable. Explanations allow us to understand how and why an event happened, which immediately allows us to see how and why it might happen again. Indeed, whenever we say that something *can't* happen – for example, mind reading or levitation or a law that limits the power of incumbents – we usually just mean that we'd have no way to explain it if it did. Unexplained events seem rare, and rare events naturally have a greater emotional impact than common events do. We are awed by a solar eclipse but merely impressed by a sunset despite the fact that the latter is by far the more spectacular visual treat.

The second reason why unexplained events have a disproportionate emotional impact is that we are especially likely to keep thinking about them. People spontaneously try to explain events,[36] and studies show that when people do not complete the things they set out to do, they are especially likely to think about and remember their unfinished business.[37] Once we explain an event, we can fold it up like freshly washed laundry, put it away in memory's drawer, and move on to the next one; but if an event defies explanation, it becomes a *mystery* or a *conundrum* – and if there's one thing we all know about mysterious conundrums, it is that they generally refuse to stay in the back of our minds. Filmmakers and novelists often capitalize on this fact by fitting their narratives with mysterious endings, and research shows that people are, in fact, more likely to keep thinking about a movie when they can't explain what happened to the main character. And if they *liked* the movie, this morsel of mystery causes them to remain happy longer.[38]

Explanation robs events of their emotional impact because it makes them seem likely and allows us to stop thinking about them. Oddly enough, an explanation doesn't actually have to *explain* anything to have these effects – it merely needs to *seem* as though it does. For instance, in one study, a researcher approached college students in the university library, handed them one of two cards with a dollar coin attached, then walked away. You'd probably agree that this is a curious event that begs for explanation. As figure 20 shows, both cards stated that the researcher was a member of the 'Smile Society', which was devoted to 'random acts of kindness'. But one card also contained two extra phrases – 'Who are we?' and 'Why do we do this?' These empty phrases didn't really provide any new information, of course, but they made students *feel* as though the curious event had been explained ('Aha, *now* I understand why they gave me a dollar!'). About five minutes later, a different researcher approached the student and claimed to be doing a class project on 'community thoughts and feelings'. The researcher asked the student to complete some survey questions, one of which was "How positive or negative are you feeling right now?" The results showed that those students who had received a card with the pseudo-explanatory phrases felt less happy than those who had received a card without them. Apparently, even a fake explanation can cause us to tuck an event away and move along to the next one.

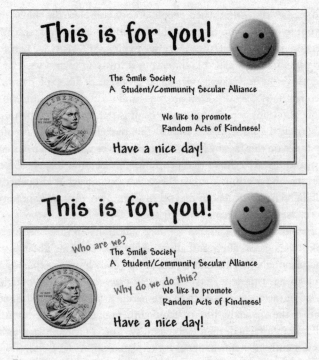

Fig. 20.

Uncertainty can preserve and prolong our happiness, thus we might expect people to cherish it. In fact, the opposite is generally the case. When a new group of students was asked which of the two cards shown in figure 20 would make them happier, 75 per cent chose the one with the meaningless explanation. Similarly, when a group of students was asked whether they would prefer to know or not know which of the simulated students had written each of the glowing reports in the online chat-room study, 100 per cent chose to know. In both cases, students chose certainty over uncertainty and clarity over mystery – despite the fact that in both cases clarity and certainty had been shown to diminish happiness. The poet John Keats noted that whereas great authors are 'capable of being in uncertainties, mysteries, doubts, without any irritable reaching after fact and reason', the rest of us are 'incapable of remaining content with half-knowledge'.[39] Our relentless desire to explain everything that happens may well distinguish us from fruit flies, but it can also kill our buzz.

Onward

The eye and the brain are conspirators, and like most conspiracies, theirs is negotiated behind closed doors, in the back room, outside of our awareness. Because we do not realize that we have generated a positive view of our current experience, we do not realize that we will do so again in the future. Not only does our naïveté cause us to overestimate the intensity and duration of our distress in the face of future adversity, but it also leads us to take actions that may undermine the conspiracy. We are more likely to generate a positive and credible view of an action than an inaction, of a painful experience than of an annoying experience, of an unpleasant situation that we cannot escape than of one we can. And yet, we rarely choose action over inaction, pain over annoyance and commitment over freedom. The processes by which we generate positive views are many: we pay more attention to favourable information, we surround ourselves with those who provide it and we accept it uncritically. These tendencies make it easy for us to explain unpleasant experiences in ways that exonerate us and make us feel better. The price we pay for our irrepressible explanatory urge is that we often spoil our most pleasant experiences by making good sense of them.

Our tour of imagination has covered a lot of ground – from realism to presentism to rationalization – so before moving on to our final destination, it may be useful to locate ourselves on the big map. We've seen how difficult it is to predict accurately our emotional reactions to future events because it is difficult to imagine them as they will happen, and difficult to imagine how we'll think about them once they do. Throughout this book, I've compared imagination to perception and memory, and I've tried to convince you that foresight is just as fallible as eyesight and hindsight. Fallible eyesight can be remedied by glasses and fallible hindsight can be remedied by written records of the past – but what of fallible foresight? There are no spectacles that can sharpen our view of tomorrow and no records of things to come. Can we remedy the problems of foresight? As we are about to see, we can. But we generally choose not to.

PART VI

Corrigibility

corrigibility (kor·i·dzĭ·b´l·ĭ·tee)
Capacity for being corrected, reformed
or improved.

CHAPTER 10

Once Bitten

Experience, O, thou disprov'st report!

Shakespeare, *Cymbeline*

THE LAST DECADE has seen an explosion of books about poo. When my two-year-old granddaughter crawls up onto my lap, she typically brings with her a fat stack of picture books, including several that explore in considerable detail the miracle of defecation and the mysteries of indoor plumbing. Some offer detailed descriptions for the budding anatomist; some offer little more than drawings of happy children, squatting, standing and wiping. Despite their many differences, each of these books communicates the same message: *grown-ups do not poo in their pants, but if you do, then don't worry too much about it.* My granddaughter seems to find this message both reassuring and inspirational. She understands that there is a right way and a wrong way to poo, and that while we don't expect her to poo the right way just yet, we do want her to notice that most of the people around her have learned to poo the right way, which suggests that with a little practice and a little coaching, she can learn to poo the right way too.

As it turns out, the benefits of practice and coaching are not limited to this particular skill. In fact, practice and coaching are the two means by which we learn just about everything we know. Firsthand knowledge and secondhand knowledge are the only two kinds of knowledge there are, and no matter what task we master – pooing, cooking, investing, bobsledding – that mastery is always a product of direct experience and/or of listening to those who have had direct experience. Babies poo in their nappies because they are rookies and because they cannot take advantage of the lessons that veterans can provide. Because babies lack both firsthand knowledge and second-hand knowledge of proper potty protocol, we expect them to make a stinky mess – but we also expect that within a few years, practice and coaching will begin to have their remedial effects, innocence will yield to experience and education, and pooing errors will disappear altogether. So why doesn't this analysis extend to errors of every kind? We all have direct experience with things that do or don't make us happy, we all have friends, therapists, cabdrivers and talk-show hosts who tell us about things that will or won't make us happy, and yet, despite all this practice and all this coaching, our search for happiness often culminates in a stinky mess. We expect the next car, the next house or the next promotion to make us happy even though the last ones didn't and even though others keep telling us that the next ones won't. Why don't we learn to avoid these mistakes in the same way that we learn to avoid warm nappies? If practice and coaching can teach us to keep our pants dry, then why can't they teach us to predict our emotional futures?

The Least Likely of Times

There are many good things about getting older, but no one knows what they are. We fall asleep and wake up at all the wrong times, avoid more foods than we can eat and take pills to help us remember which other pills to take. In fact, the only really good thing about getting older is that people who still have all their hair are occasionally forced to stand back and admire our wealth of experience. They think of our experience as a form of wealth because they assume it allows us to avoid making the same mistake twice – and sometimes it does. There *are* a few experiences that those of us who are filthy rich with it just don't repeat, and bathing a cat while drinking peppermint schnapps comes to mind for reasons I'd rather not discuss right now. On the other hand, there are plenty of mistakes that we highly experienced people seem to make over and over again. We marry people who are oddly like the people we divorced, we attend annual family gatherings and make an annual vow never to return, and we carefully time our monthly expenditures to ensure that we will once again be flat broke on all the dates that begin with a three. These cycles of recidivism seem difficult to explain. After all, shouldn't we learn from our own experience? Imagination has its shortcomings, to be sure, and it is perhaps inevitable that we will mispredict how future events will make us feel when we've never experienced those events before. But once we've been married to a busy executive who spends more hours at work than at home, once we've attended a family reunion at which the aunts fight with the uncles who do their best to offend the cousins, and once we've spent a few lean days between paycheques acquiring intimate knowledge of rice and beans, shouldn't we be able to imagine these events with a reasonable degree of accuracy and hence take steps to avoid them in the future?

We should and we do, but not as often or as well as you might expect. We *try* to repeat those experiences that we remember with pleasure and pride, and we *try* to avoid repeating those that we remember with embarrassment and regret.[1] The trouble is that we often don't remember them correctly. Remembering an experience *feels* a lot like opening a drawer and retrieving a story that was filed away on the day it was written, but as we've seen in previous chapters, that feeling is one of our brain's most sophisticated illusions. Memory is not a dutiful scribe that keeps a complete transcript of our experiences, but a sophisticated editor that clips and saves key elements of an experience and then uses these elements to rewrite the story each time we ask to reread it. The clip-and-save method usually works pretty well because the editor usually has a keen sense of which elements are essential and which are disposable. That's why we remember how the groom looked when he kissed the bride but not which finger the flower girl had up her nose when it happened. Alas, as keen as its editorial skills may be, memory does have a few quirks that cause it to misrepresent the past, and this causes us to misimagine the future.

For example, you may or may not *use* four-letter words, but I trust you've never counted them. So take a guess: are there more four-letter words in the English language that begin with *k* (k-1's) or that have *k* as their third letter (k-3's)? If you are like most people, you guessed that the k-1's outnumber the k-3's.[2] You probably answered this question by briefly checking your memory ('Hmmm, there's *kite, kilt, kale* . . . '), and because you found it easier to recall k-1's than k-3's, you assumed there must *be* more of the former than the latter. This would normally be a very fine deduction. After all, you can *recall* more four-legged elephants (e-4's) than six-legged elephants (e-6's) because you have *seen* more e-4's than e-6's, and you have *seen* more e-4's than e-6's because there *are* more e-4's than e-6's. The actual number of e-4's and e-6's in the world determines how frequently you encounter them, and the frequency of your encounters determines how easily you can remember those encounters.

Alas, the reasoning that serves you so well when it comes to elephants serves you quite poorly when it comes to words. It is indeed easier to recall k-1's than k-3's, but *not* because you have encountered more of the former than the latter. Rather, it's easier to recall words that start with *k* because it is easier to recall *any* word by its first letter than by its third letter. Our mental dictionaries are organized more or less alphabetically, like Webster's itself, hence we can't easily "look up" a word in our memories by any letter except the first one. The fact is that there are many more k-3's than k-1's in the English language, but because the latter are easier to recall, people routinely get this question wrong. The *k*-word puzzle works because we naturally (but incorrectly) assume that things that come easily to mind are things we have frequently encountered.

What is true of elephants and words is also true of experiences.[3] Most of us can bring to mind memories of riding a bicycle more easily than we can bring to mind memories of riding a yak, hence we correctly conclude that we've ridden more bikes than yaks in the past. This would be impeccable logic – except for the fact that the frequency with which we've had an experience is not the only determinant of the ease with which we remember it. In fact, infrequent or *unusual* experiences are often among the most memorable, which is why most Americans know precisely where they were on the morning of 11 September 2001, but not on the morning of 10 September.[4] The fact that infrequent experiences come so readily to mind can lead us to draw some peculiar conclusions. For instance, for most of my adult life I have had the distinct impression that I tend to pick the slowest queue at the grocery store, and that whenever I get tired of waiting in the slowest queue and switch to another, the queue I switched from begins moving faster than the one I switched to.[5] Now, if this were true – if I really did have bad karma, bad juju or some other metaphysical form of badness that caused any queue I joined to slow down – then there would have to be someone out there who felt that they had a metaphysical form of goodness that caused any queue they joined to speed up. After all, *everyone* can't get in the slowest queue on every occasion, can they? And yet, nobody *I* know feels that they have the power to quicken queues by joining them. On the contrary, just about everyone I know seems to believe that they, like me, are inexorably drawn to the slowest of all possible queues, and that their occasional attempts to thwart fate merely slow the queues they join and hasten the queues they abandon. Why do we all believe this?

Because standing in a queue that is moving at a rapid pace, or even an average pace, is such a mind-numbingly ordinary experience that we don't notice or remember it. Instead we just stand there bored, glancing at the tabloids, contemplating the chocolate bars and wondering what idiot decided that batteries of different sizes should be labelled with different numbers of *A*'s rather than with words we can actually remember such as *large*, *medium* and *small*. As we do this, we rarely turn to our partners and say, "Have you noticed how *normally* this queue is moving? I mean, it's just so damned average that I'm feeling compelled to make notes so that I can charm others with the tale at a later date." No, the queue-moving experiences we *remember* are those in which the guy in the bright red hat who was originally standing behind us before he switched to the other queue has made it out of the store and into his car before we've even made it to the cash register because the bovine grandmother ahead of us is waving her coupons at the clerk and debating the true meaning of the phrase *expiration date*. This doesn't really happen that often, but because it is so memorable, we tend to think it does.

The fact that the *least likely experience* is often the *most likely memory* can wreak havoc with our ability to predict future experiences.[6] For example, in one study, researchers asked commuters waiting on a subway platform to imagine how they would feel if they missed their train that day.[7] Before making this prediction, some of the commuters (any-memory group) were asked to remember and describe 'a time you missed your train'. Other commuters (worst-memory group) were asked to remember and describe 'the *worst* time you missed your train'. The results showed that commuters in the any-memory group remembered an episode that was every bit as awful as the episode remembered by commuters in the worst-memory group. In other words, when commuters thought about train-missing, the single most inconvenient and frustrating episodes tended to come to their minds ('I could hear the train arriving and so I started to run to catch it, but I tripped on the stairs and knocked over this guy who was selling umbrellas, and as a result I was a half hour late for a job interview and by the time I got there they had already hired someone else'). Most instances of train missing are ordinary and forgettable, hence when we think about train missing, we tend to remember the most extraordinary instances.

Now, what does this have to do with predicting our emotional futures? K-1 words come quickly to mind because of the way our mental dictionaries are organized and not because they are common, and memories of slow-moving supermarket queues come quickly to mind because we pay special attention when we're stuck in them and not because they are common. But because we don't recognize the *real* reasons why these memories come quickly to mind, we mistakenly conclude that they are more common than they actually are. Similarly, awful train-missing incidents come quickly to mind not because they are common but because they are *uncommon*. But because we don't recognize the *real* reasons why these awful episodes come quickly to mind, we mistakenly conclude that they are more common than they actually are. And indeed, when commuters were asked to make *predictions* about how they would feel if they missed their train *that day*, they mistakenly expected the experience to be much more inconvenient and frustrating than it likely would have been.

This tendency to recall and rely on unusual instances is one of the reasons why we so often repeat mistakes. When we think about last year's family holiday we do not recruit a fair and representative sample of instances from our two-week tour of Idaho. Instead, the memory that comes most naturally and quickly to mind is of that first Saturday afternoon when we took the kids horseback riding, crested the ridge on our palominos and found ourselves looking down into a magnificent valley, the river wending its way to the horizon like a mirrored ribbon as the sun played on its surface. The air was crisp, the woods were quiet. The kids suddenly stopped arguing and sat transfixed on their horses, someone said "Wow" in a very soft voice, everyone smiled at everyone else, and the moment was forever crystallized as the high point of the holiday. Which is why it instantly springs to mind. But if we rely on this memory as we plan our next holiday while overlooking the fact that the rest of the trip was generally disappointing, we risk finding ourselves at the same overcrowded campground the next year, eating the same stale sandwiches, being bitten by the same surly ants and wondering how we managed to learn so little from our previous visit. Because we tend to remember the best of times and the worst of times instead of the most likely of times, the wealth of experience that young people admire does not always pay clear dividends.

All's Well

I recently had an argument with my wife, who insisted that I like the movie *Schindler's List*. Now, let me be clear: she was not insisting that I *would* like the film or that I *should* like the film. She was insisting that I *do* like the film, which we saw together in 1993. This struck me as supremely unfair. I don't get to be right about too many things, but the one thing I reserve the right to be right about is what I like. And as I have been telling everyone who would listen for more than a decade, I do not like *Schindler's List*. But my wife said I was wrong, and as a scientist I feel morally bound to test any hypothesis that involves me eating popcorn. So we rented *Schindler's List*, watched it again, and the results of my experiment unequivocally proved who was right: we were. She was right because I was indeed riveted by the movie for all of the first two hundred minutes. But I was right because at the end something awful happened. Rather than leaving me at the story's conclusion, the director, Steven Spielberg, added a final scene in which the real people on whom the characters were based came on-screen and honored the movie's hero. I found that scene so intrusive, so mawkish, so thoroughly superfluous, that I actually said to my wife, 'Oh, give me a break', which is apparently what I'd said in a rather loud voice to the entire theatre in 1993. The first 98 per cent of the movie was brilliant, the final 2 per cent was stupid, and I remembered not liking the movie because (for me) it had ended badly. The only strange thing about this memory is that I've sat through an awful lot of films whose proportion of brilliance was significantly less than 98 per cent, and I remember liking some of them quite a bit. The difference is that in those films the stupid parts were at the beginning, or in the middle, or somewhere other than the very end. So why do I like average films that end superbly more than nearly perfect films that end badly? After all, don't I get more minutes of intense and satisfying emotional involvement with the nearly perfect film than with the average film?

Yes, but apparently that's not what matters. As we've seen, memory does not store a feature-length film of our experience but instead stores an idiosyncratic synopsis, and among memory's idiosyncrasies is its obsession with final scenes.[8] Whether we hear a series of sounds, read a series of letters, see a series of pictures, smell a series of odours, or meet a series of people, we show a pronounced tendency to recall the items at the end of the series far better than the items at the beginning or in the middle.[9] As such, when we look back on the entire series, our impression is strongly influenced by its final items.[10] This tendency is particularly acute when we look back on experiences of pleasure and pain. For instance, volunteers in one study were asked to submerge their hands in icy water (a common laboratory task that is quite painful but that causes no harm) while using an electronic rating scale to report their moment-to-moment discomfort.[11] Every volunteer performed both a short trial and a long trial. On the short trial, the volunteers submerged their hand for sixty seconds in a water bath that was kept at a chilly fifty-seven degrees Fahrenheit. On the long trial, volunteers submerged their hand for ninety seconds in a water bath that was kept at a chilly fifty-seven degrees Fahrenheit for the first sixty seconds, then surreptitiously warmed to a not-quite-as-chilly fifty-nine degrees over the remaining thirty seconds. So the short trial consisted of sixty cold seconds, and the long trial consisted of the *same sixty cold seconds* with *an additional thirty cool seconds*. Which trial was more painful?

Well, it depends on what we mean by *painful*. The long trial clearly comprised a greater number of painful moments, and indeed, the volunteers' moment-to-moment reports revealed that they experienced equal discomfort for the first sixty seconds on both trials, but much more discomfort in the next thirty seconds if they kept their hand in the water (as they did on the long trial) than if they removed it (as they did on the short trial). On the other hand (sorry), when volunteers were later asked to *remember* their experience and say which trial had *been* more painful, they tended to say that the short trial had been more painful than the long one. Although the long trial required the volunteers to endure 50 per cent more seconds of immersion in ice water, it had a slightly warmer finish and hence was remembered as the less painful of the two experiences. Memory's fetish for endings explains why women often remember childbirth as less painful than it actually was,[12] and why couples whose relationships have gone sour remember that they were never really happy in the first place.[13] As Shakespeare wrote, 'The setting sun, and music at the close / As the last taste of sweets, is sweetest last / Writ in remembrance more than things long past.'[14]

The fact that we often judge the pleasure of an experience by its ending can cause us to make some curious choices. For example, when the researchers who performed the cold-water study asked the volunteers which of the two trials they would prefer to repeat, 69 per cent of the volunteers chose to repeat the long one – that is, *the one that entailed an extra thirty seconds of pain*. Because the volunteers remembered the long trial as less painful than the short one, that was the one they chose to repeat. It would be easy to impugn the rationality of this choice – after all, the 'total pleasure' of an experience is a function of both the quality and the quantity of the moments that constitute it, and these volunteers were clearly not considering quantity.[15] But it would be just as easy to defend the rationality of this choice. We don't ride the mechanical bull or pose for a picture with a handsome movie star because these momentary experiences are inherently pleasurable; we do it so that we can spend the rest of our years immersed in blissful recollection ('I stayed on for a full minute!'). If we can spend hours enjoying the memory of an experience that lasted just a few seconds, and if memories tend to overemphasize endings, then why not endure a little extra pain in order to have a memory that is a little less painful?[16]

Both of these positions are sensible, and you could sensibly hold either of them. The problem is that you more than likely hold *both* of them. Consider, for example, a study in which volunteers learned about a woman (whom we'll call Ms Dash) who had an utterly fabulous life until she was sixty years old, at which point her life went from utterly fabulous to merely satisfactory.[17] Then, at the age of sixty-five, Ms Dash was killed in an auto accident. How good was her life (which is depicted by the dashed line in figure 21)? On a nine-point scale, volunteers said that Ms Dash's life was a 5.7. A second group of volunteers learned about a woman (whom we'll call Ms Solid) who had an utterly fabulous life until she was killed in an auto accident at the age of sixty. How good was her life (which is depicted by the solid line in figure 21)? Volunteers said that Ms Solid's life was a 6.5. It appears, then, that these volunteers preferred a fabulous life (Ms Solid's) to an equally fabulous life with a few additional merely satisfactory years (Ms Dash's). If you think about it for a moment, you'll realize that this is just how the volunteers in the ice-water study were thinking. Ms Dash's life had more 'total pleasure' than did Ms Solid's, Ms Solid's life had a better ending than Ms Dash's, and the volunteers were clearly more concerned with the quality of a life's ending than with the total quantity of pleasure the life contained. But wait a minute. When a third group of volunteers was asked to compare the two lives side by side (as you can do at the bottom of figure 21), they showed no such preference. When the difference in the quality of the two lives was made

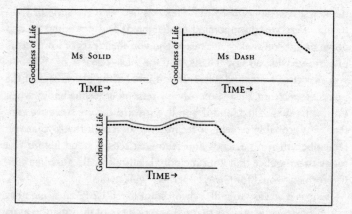

Fig. 21. When considered separately, the shapes of the curves matter. But when directly compared, the lengths of the curves matter.

salient by asking volunteers to consider them simultaneously, the volunteers were no longer so sure that they preferred to live fast, die young and leave a happy corpse. Apparently, the way an experience ends is more important to us than the total amount of pleasure we receive – until we think about it.

The Way We Weren't

If you were an American of voting age on the evening of 8 November 1988, then you were probably at home watching the results of the presidential contest between Michael Dukakis and George Bush. When you think back on that election you may recall the infamous Willie Horton ad, or the phrase 'card-carrying member of the ACLU', or Lloyd Bentsen's stinging retort to Dan Quayle: 'Senator, you're no Jack Kennedy.' You most certainly recall that when all the votes were tallied, Americans decided not to send a Massachusetts liberal to the White House. Although Dukakis lost the election, he did win some of the more liberal states, and because we're talking about memory, I'd like to ask you to use yours right now. Close your eyes for a moment and try to remember exactly how you felt when the newscaster announced that Dukakis had won the state of California. Were you disappointed or delighted? Did you jump up and down or did you shake your head? Did you shed tears of joy or tears of sorrow? Did you say, 'Thank God for the left coast!' or 'What do you expect of fruits and nuts?' If you inhabit the liberal end of the political spectrum, then you probably remember feeling happy when California was called, and if you live toward the conservative end, then you probably remember feeling less so. And if that's what you remember, then my friends and fellow citizens, I stand before you today to announce that your memory is mistaken. Because in 1988, Californians voted for George Herbert Walker Bush.

Why is it so easy to pull a cheap stunt like this? Because memory is a reconstructive process that uses every piece of information at its disposal to build the mental images that come trippingly to mind when we engage in an act of remembering. One such piece of information is the fact that California is the liberal state that gave us Transcendental Meditation, muesli, psychedelic rock, Governor Moonbeam and *Debbie Does Dallas*. So it makes sense that Michael Dukakis – like Bill Clinton, Al Gore and John Kerry – would have won the state handily. But before Californians began voting for Bill Clinton, Al Gore and John Kerry, they voted as many times as they possibly could for Gerald Ford, Ronald Reagan and Richard Nixon. Unless you are a political scientist, a CNN junkie or a longtime Californian, you probably didn't remember that bit of historical politrivia. Instead you made a logical inference, namely, that because California is a liberal state, and because Dukakis was a liberal candidate, Californians must have voted for him. Just as anthropologists use both facts (a thirteen-thousand-year-old skull found near Mexico City is long and narrow) and theories ('Long, narrow skulls indicate European ancestry') to make guesses about past events ('Caucasians came to the New World two thousand years before the Mongoloids replaced them'), so did your brain use a fact ('Dukakis was a liberal') and a theory ('Californians are liberal') to make a guess about a past event ('Californians voted for Dukakis'). Alas, because your theory was wrong, your guess was wrong too.

Our brains use facts and theories to make guesses about past events, and so too do they use facts and theories to make guesses about past feelings.[18] Because feelings do not leave behind the same kinds of facts that presidential elections and ancient civilizations do, our brains must rely even more heavily on theories to construct memories of how we once felt. When those theories are wrong, we end up misremembering our own emotions. Consider, for instance, how your theories about something – oh, say, how about gender? – might influence your recollection of past feelings. Most of us believe that men are less emotional than women ('She cried, he didn't'), that men and women have different emotional reactions to similar events ('He was angry, she was sad'), and that women are particularly prone to negative emotions at particular points in their menstrual cycles ('She's a bit irritable today, if you know what I mean'). As it turns out, there is little evidence for any of these beliefs – but that's not the point. The point is that these beliefs are theories that can influence how we remember our own emotions. Consider:

- In one study, volunteers were asked to remember how they had felt a few months earlier, and the male and female volunteers remembered feeling equally intense emotions.[19] Another group of volunteers was asked to remember how they had felt a month earlier, but before doing so, they were asked to think a bit about gender. When volunteers were prompted to think about gender, female volunteers remembered feeling more intense emotion and male volunteers remembered feeling less intense emotion.

- In one study, male and female volunteers became members of teams and played a game against an opposing team.[20] Some volunteers immediately reported the emotions they had felt while playing the game, and others recalled their emotions a week later. Male and female volunteers did not differ in the kinds of emotions they reported. But a week later female volunteers recalled feeling more stereotypically feminine emotions (e.g., sympathy and guilt) and male volunteers recalled feeling more stereotypically masculine emotions (e.g., anger and pride).

- In one study, female volunteers kept diaries and made daily ratings of their feelings for four to six weeks.[21] These ratings revealed that women's emotions did not vary with the phase of their menstrual cycles. However, when the women were later asked to reread the diary entry for a particular day and remember how they had been feeling, they remembered feeling more negative emotion on the days on which they were menstruating.

It seems that our theories about how people of our gender *usually* feel can influence our memory of how we actually felt. Gender is but one of many theories that have this power to alter our memories. For instance, Asian culture does not emphasize the importance of personal happiness as much as European culture does, and thus Asian Americans believe that they are generally less happy than their European American counterparts. In one study, volunteers carried handheld computers everywhere they went for a week and recorded how they were feeling when the computer beeped at random intervals throughout the day.[22] These reports showed that the Asian American volunteers were slightly happier than the European American volunteers. But when the volunteers were asked to *remember* how they had felt that week, the Asian American volunteers reported that they had felt *less* happy and not more. In a study using similar methodology, Hispanic Americans and European Americans reported feeling pretty much the same during a particular week, but the Hispanic Americans remembered feeling happier than the European Americans did.[23] Not all theories involve some immutable characteristic of persons, such as gender or culture. For example, which students tend to score highest on an exam – those who worry about grades, or those who don't? As a college professor, I can tell you that my own theory is that students who are deeply concerned about their performance tend to study more and hence outscore their more lackadaisical classmates. Apparently students have the same theory, because research shows that when students do well on an exam, they remember feeling more anxious before the exam than they actually felt, and when students do poorly on an exam, they remember feeling less anxious before the exam than they actually felt.[24]

We remember feeling as we believe we must have felt. The problem with this error of retrospection is that it can keep us from discovering our errors of prospection. Consider the case of the 2000 US presidential election. Voters went to the polls on 7 November 2000, to decide whether George Bush or Al Gore would become the forty-third president of the United States, but it quickly became clear that the election was too close to call and that its outcome would take weeks to decide. The next day, 8 November researchers asked some voters to predict how happy they would be on the day the election was ultimately decided for or against their favoured candidate. On 13 December Al Gore conceded to George Bush, and the next day, 14 December the researchers measured the actual happiness of the voters. Four months later, in April 2001, the researchers contacted the voters again and asked them to recall how they had felt on 14 December. As figure 22 shows, the study revealed three things. First,

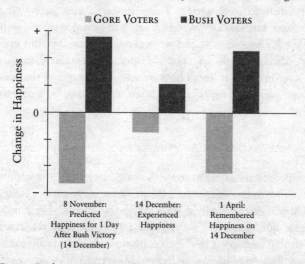

Fig. 22. In the 2000 US presidential election, partisans expected the Supreme Court's decision to strongly influence how happy they would feel a day after the decision was announced *(bars on the left).* A few months later they remembered that it had *(bars on the right).* In fact, the decision had a far smaller impact on happiness than the partisans either predicted or remembered *(bars in the middle).*

on the day after the election, pro-Gore voters expected to be devastated and pro-Bush voters expected to be elated if George Bush was ultimately declared the winner. Second, when George Bush was ultimately declared the winner, pro-Gore voters were less devastated and pro-Bush voters were less elated than they had expected to be (a tendency you've seen before in other chapters). But third and most important, a few months after the election was decided, both groups of voters remembered feeling as they *had expected* to feel, and not as they *had actually* felt. Apparently, prospections and retrospections can be in perfect agreement despite the fact that neither accurately describes our actual experience.[25] The theories that lead us to predict that an event will make us happy ("If Bush wins, I'll be elated") also lead us to remember that it did ("When Bush won, I was elated"), thereby eliminating evidence of their own inaccuracy. This makes it unusually difficult for us to discover that our predictions were wrong. We overestimate how happy we will be on our birthdays,[26] we underestimate how happy we will be on Monday mornings,[27] and we make these mundane but erroneous predictions again and again, despite their regular disconfirmation. Our inability to recall how we really felt is one of the reasons why our wealth of experience so often turns out to be a poverty of riches.

Onward

When people are asked to name the single object they would try to save if their home caught fire, the most common answer (much to the chagrin of the family dog) is 'My photo album'. We don't just treasure our memories; we *are* our memories. And yet, research reveals that memory is less like a collection of photographs than it is like a collection of impressionist paintings rendered by an artist who takes considerable licence with his subject. The more ambiguous the subject is, the more licence the artist takes, and few subjects are more ambiguous than emotional experience. Our memory for emotional episodes is overly influenced by unusual instances, closing moments and theories about how we *must* have felt way back then, all of which gravely compromise our ability to learn from our own experience. Practice, it seems, doesn't always make perfect. But if you think back to all that talk about pooing, you'll remember that practice is just one of the two ways in which we learn. If practice doesn't fix us, then what about coaching?

CHAPTER 11

Reporting Live from Tomorrow

> Instructed by the antiquary times,
> He must, he is, he cannot but be wise.
>
> Shakespeare, *Troilus and Cressida*

IN ALFRED HITCHCOCK'S 1956 REMAKE of *The Man Who Knew Too Much*, Doris Day sang a waltz whose final verse went like this:

> When I was just a child in school,
> I asked my teacher, "What will I try?
> Should I paint pictures, should I sing songs?"
> This was her wise reply:
> '*Que sera, sera.* Whatever will be, will be.
> The future's not ours to see. *Que sera, sera.*'[1]

Now, I don't mean to quibble with the lyricist, and I have nothing but fond memories of Doris Day, but the fact is that this is *not* a particularly wise reply. When a child asks for advice about which of two activities to pursue, a teacher should be able to provide more than a musical cliché. Yes, of *course* the future is hard to see. But we're all heading that way anyhow, and as difficult as it may be to envision, we have to make *some* decisions about which futures to aim for and which to avoid. If we are prone to mistakes when we try to imagine the future, then how *should* we decide what to do?

Even a child knows the answer to that one: we should ask the teacher. One of the benefits of being a social and linguistic animal is that we can capitalize on the experience of others rather than trying to figure everything out for ourselves. For millions of years, human beings have conquered their ignorance by dividing the labour of discovery and then communicating their discoveries to one another, which is why the average newspaper boy in Pittsburgh knows more about the universe than did Galileo, Aristotle, Leonardo or any of those other guys who were so smart they only needed one name. We all make ample use of this resource. If you were to write down everything you know and then go back through the list and make a check mark next to the things you know only because somebody told you, you'd develop a repetitive-motion disorder because almost *everything* you know is secondhand. Was Yury Gagarin the first man in space? Is *croissant* a French word? Are there more Chinese than North Dakotans? Does a stitch in time save nine? Most of us know the answers to these questions despite the fact that none of us actually witnessed the launching of *Vostok I*, personally supervised the evolution of language, hand-counted all the people in Beijing and Bismarck or performed a fully randomized double-blind study of stitching. We know the answers because someone shared them with us. Communication is a kind of 'vicarious observation'[2] that allows us to learn about the world without ever leaving the comfort of our armchairs. The six billion interconnected people who cover the surface of our planet constitute a leviathan with twelve billion eyes, and anything that is seen by one pair of eyes can potentially be known to the entire beast in a matter of months, days or even minutes.

The fact that we can communicate with one another about our experiences should provide a simple solution to the core problem with which this book has been concerned. Yes, our ability to imagine our future emotions is flawed – but that's okay, because we don't have to imagine what it would feel like to marry a lawyer, move to Texas or eat a snail when there are so many people who have *done* these things and are all too happy to tell us about them. Teachers, neighbours, coworkers, parents, friends, lovers, children, uncles, cousins, coaches, cabdrivers, bartenders, hairstylists, dentists, advertisers – each of these people have something to say about what it would be like to live in this future rather than that one, and at any point in time we can be fairly sure that one of these people have actually *had* the experience that we are merely contemplating. Because we are the mammal that shows and tells, each of us has access to information about almost any experience we can possibly imagine – and many that we can't. Guidance counsellors tell us about the best careers, critics tell us about the best restaurants, travel agents tell us about the best holidays and friends tell us about the best travel agents. Every one of us is surrounded by a platoon of Dear Abbys who can recount their own experiences and in so doing tell us which futures are most worth wanting.

Given the overabundance of consultants, role models, gurus, mentors, matchmakers and nosy relatives, we might expect people to do quite well when it comes to making life's most important decisions, such as where to live, where to work and whom to marry. And yet, the average American moves more than six times,[3] changes jobs more than ten times[4] and marries more than once,[5] which suggests that most of us are making more than a few poor choices. If humanity is a living library of information about what it feels like to do just about anything that can be done, then why do the people with the library cards make so many bad decisions? There are just two possibilities. The first is that a lot of the advice we receive from others is bad advice that we foolishly accept. The second is that a lot of the advice we receive from others is good advice that we foolishly reject. So which is it? Do we listen too well when others speak, or do we not listen well enough? As we shall see, the answer to that question is *yes*.

Super-replicators

The philosopher Bertrand Russell once claimed that believing is 'the most mental thing we do'.[6] Perhaps, but it is also the most *social* thing we do. Just as we pass along our genes in an effort to create people whose faces look like ours, so too do we pass along our beliefs in an effort to create people whose minds think like ours. Almost any time we tell anyone anything, we are attempting to change the way their brains operate – attempting to change the way they see the world so that their view of it more closely resembles our own. Just about every assertion – from the sublime ('God has a plan for you') to the mundane ('Turn left at the light, go two miles, and you'll see the Dunkin' Donuts on your right') – is meant to bring the listener's beliefs about the world into harmony with the speaker's. Sometimes these attempts succeed and sometimes they fail. So what determines whether a belief will be successfully transmitted from one mind to another?

The principles that explain why some genes are transmitted more successfully than others also explain why some beliefs are transmitted more successfully than others.[7] Evolutionary biology teaches us that any gene that promotes its own 'means of transmission' will be represented in increasing proportions in the population over time. For instance, imagine that a single gene were responsible for the complex development of the neural circuitry that makes orgasms feel so good. For a person having this gene, orgasms would feel . . . well, orgasmic. For a person lacking this gene, orgasms would feel more like sneezes – brief, noisy, physical convulsions that pay rather paltry hedonic dividends. Now, if we took fifty healthy, fertile people who had the gene and fifty healthy, fertile people who didn't, and left them on a hospitable planet for a million years or so, when we returned we would probably find a population of thousands or millions of people, almost all of whom had the gene. Why? Because a gene that made orgasms feel good would tend to be transmitted from generation to generation simply because people who enjoy orgasms are inclined to do the thing that transmits their genes. The logic is so circular that it is virtually inescapable: genes tend to be transmitted when they make us do the things that transmit genes. What's more, even *bad* genes – those that make us prone to cancer or heart disease – can become super-replicators if they compensate for these costs by promoting their own means of transmission. For instance, if the gene that made orgasms feel delicious also left us prone to arthritis and tooth decay, that gene might still be represented in increasing proportions because arthritic, toothless people who love orgasms are more likely to have children than are limber, toothy people who do not.

The same logic can explain the transmission of beliefs. If a particular belief has some property that facilitates its own transmission, then that belief tends to be held by an increasing number of minds. As it turns out, there are several such properties that increase a belief's transmissional success, the most obvious of which is accuracy. When someone tells us where to find a parking space downtown or how to bake a cake at high altitude, we adopt that belief and pass it along because it helps us and our friends do the things we want to do, such as parking and baking. As one philosopher noted, 'The faculty of communication would not gain ground in evolution unless it was by and large the faculty of transmitting true beliefs.'[8] Accurate beliefs give us power, which makes it easy to understand why they are so readily transmitted from one mind to another.

It is a bit more difficult to understand why *inaccurate* beliefs are so readily transmitted from one mind to another – but they are. False beliefs, like bad genes, can and do become super-replicators, and a thought experiment illustrates how this can happen. Imagine a game that is played by two teams, each of which has a thousand players, each of whom is linked to teammates by a telephone. The object of the game is to get one's team to share as many accurate beliefs as possible. When players receive a message that they believe to be accurate, they call a teammate and pass it along. When they receive a message that they believe to be inaccurate, they don't. At the end of the game, the referee blows a whistle and awards each team a point for every accurate belief that the entire team shares and subtracts one point for every inaccurate belief the entire team shares. Now, consider a contest played one sunny day between a team called the Perfects (whose members always transmit accurate beliefs) and a team called the Imperfects (whose members occasionally transmit an inaccurate belief). We should expect the Perfects to win, right?

Not necessarily. In fact, there are some special circumstances under which the Imperfects will beat their pants off. For example, imagine what would happen if one of the Imperfect players sent the false message 'Talking on the phone all day and night will ultimately make you very happy', and imagine that other Imperfect players were gullible enough to believe it and pass it on. This message is inaccurate and thus will cost the Imperfects a point in the end. But it may have the compensatory effect of keeping more of the Imperfects on the telephone for more of the time, thus increasing the total number of accurate messages they transmit. Under the right circumstances, the costs of this inaccurate belief would be outweighed by its benefits, namely, that it led players to behave in ways that increased the odds that they would share other accurate beliefs. The lesson to be learned from this game is that inaccurate beliefs can prevail in the belief-transmission game if they somehow facilitate their own 'means of transmission'. In this case, the means of transmission is not sex but communication, and thus any belief – even a false belief – that increases communication has a good chance of being transmitted over and over again. False beliefs that happen to promote stable societies tend to propagate because people who hold these beliefs tend to live in stable societies, which provide the means by which false beliefs propagate.

Some of our cultural wisdom about happiness looks suspiciously like a super-replicating false belief. Consider money. If you've ever tried to sell anything, then you probably tried to sell it for as much as you possibly could, and other people probably tried to buy it for as little as they possibly could. All the parties involved in the transaction assumed that they would be better off if they ended up with more money rather than less, and this assumption is the bedrock of our economic behavior. Yet, it has far fewer scientific facts to substantiate it than you might expect. Economists and psychologists have spent decades studying the relation between wealth and happiness, and they have generally concluded that wealth increases human happiness when it lifts people out of abject poverty and into the middle class but that it does little to increase happiness thereafter.[9] Americans who earn $50,000 per year are much happier than those who earn $10,000 per year, but Americans who earn $5 million per year are not much happier than those who earn $100,000 per year. People who live in poor nations are much less happy than people who live in moderately wealthy nations, but people who live in moderately wealthy nations are not much less happy than people who live in extremely wealthy nations. Economists explain that wealth has 'declining marginal utility', which is a fancy way of saying that it hurts to be hungry, cold, sick, tired and scared, but once you've bought your way out of these burdens, the rest of your money is an increasingly useless pile of paper.[10]

So once we've earned as much money as we can actually enjoy, we quit working and enjoy it, right? Wrong. People in wealthy countries generally work long and hard to earn more money than they can ever derive pleasure from.[11] This fact puzzles us less than it should. After all, a rat can be motivated to run through a maze that has a cheesy reward at its end, but once the little guy is all topped up, then even the finest Stilton won't get him off his haunches. Once we've eaten our fill of pancakes, more pancakes are not rewarding, hence we stop trying to procure and consume them. But not so, it seems, with money. As Adam Smith, the father of modern economics, wrote in 1776: 'The desire for food is limited in every man by the narrow capacity of the human stomach; but the desire of the conveniences and ornaments of building, dress, equipage, and household furniture, seems to have no limit or certain boundary.'[12]

If food and money both stop pleasing us once we've had enough of them, then why do we continue to stuff our pockets when we would not continue to stuff our faces? Adam Smith had an answer. He began by acknowledging what most of us suspect anyway, which is that the production of wealth is not necessarily a source of personal happiness.

> In what constitutes the real happiness of human life, [the poor] are in no respect inferior to those who would seem so much above them. In ease of body and peace of mind, all the different ranks of life are nearly upon a level, and the beggar, who suns himself by the side of the motorway, possesses that security which kings are fighting for.[13]

That sounds lovely, but if it's true, then we're all in big trouble. If rich kings are no happier than poor beggars, then why should poor beggars stop sunning themselves by the roadside and work to become rich kings? If no one wants to be rich, then we have a significant economic problem, because flourishing economies require that people continually procure and consume one another's goods and services. Market economies require that we all have an insatiable hunger for *stuff*, and if everyone were content with the stuff they had, then the economy would grind to a halt. But if this is a significant *economic* problem, it is not a significant *personal* problem. The chair of the Federal Reserve may wake up every morning with a desire to do what the economy wants, but most of us get up with a desire to do what *we* want, which is to say that the fundamental needs of a vibrant economy and the fundamental needs of a happy individual are not necessarily the same. So what motivates people to work hard every day to do things that will satisfy the economy's needs but not their own? Like so many thinkers, Smith believed that people want just one thing – happiness – hence economies can blossom and grow only if people are deluded into believing that the production of wealth will make them happy.[14] If and only if people hold this false belief will they do enough producing, procuring and consuming to sustain their economies.

> The pleasures of wealth and greatness . . . strike the imagination as something grand and beautiful and noble, of which the attainment is well worth all the toil and anxiety which we are so apt to bestow upon it. . . . It is this deception which rouses and keeps in continual motion the industry of mankind. It is this which first prompted them to cultivate the ground, to build houses, to found cities and commonwealths, and to invent and improve all the sciences and arts, which ennoble and embellish human life; which have entirely changed the whole face of the globe, have turned the rude forests of nature into agreeable and fertile plains, and made the trackless and barren ocean a new fund of subsistence, and the great high road of communication to the different nations of the earth.[15]

In short, the production of wealth does not necessarily make individuals happy, but it does serve the needs of an economy, which serves the needs of a stable society, which serves as a network for the propagation of delusional beliefs about happiness and wealth.

Economies thrive when individuals strive, but because individuals will only strive for their own happiness, it is essential that they mistakenly believe that producing and consuming are routes to personal well-being. Although words such as *delusional* may seem to suggest some sort of shadowy conspiracy orchestrated by a small group of men in dark suits, the belief-transmission game teaches us that the propagation of false beliefs does not require that anyone be *trying* to perpetrate a magnificent fraud on an innocent populace. There is no cabal at the top, no star chamber, no master manipulator whose clever programme of indoctrination and propaganda has duped us all into believing that money can buy us love. Rather, this particular false belief is a super-replicator because holding it causes us to engage in the very activities that perpetuate it.[16]

The belief-transmission game explains why we believe some things about happiness that simply aren't true. The joy of money is one example. The joy of children is another that for most of us hits a bit closer to home. Every human culture tells its members that having children will make them happy. When people think about their offspring – either imagining future offspring or thinking about their current ones – they tend to conjure up images of cooing babies smiling from their bassinets, adorable toddlers running higgledy-piggledy across the lawn, handsome boys and gorgeous girls playing trumpets and tubas in the school marching band, successful college students going on to have beautiful weddings, satisfying careers and flawless grandchildren whose affections can be purchased with sweets. Prospective parents know that nappies will need changing, that homework will need doing and that orthodontists will go to Aruba on their life savings, but by and large, they think quite happily about parenthood, which is why most of them eventually leap into it. When parents look back on parenthood, they remember feeling what those who are looking forward to it expect to feel. Few of us are immune to these cheery contemplations. I have a twenty-nine-year-old son, and I am absolutely convinced that he is and always has been one of the greatest sources of joy in my life, having only recently been eclipsed by my two-year-old granddaughter, who is equally adorable but who has not yet asked me to walk behind her and pretend we're unrelated. When people are asked to identify their sources of joy, they do just what I do: they point to their kids.

Yet if we measure the *actual* satisfaction of people who have children, a very different story emerges. As figure 23 shows, couples generally start out quite happy in their marriages and then become progressively less satisfied over the course of their lives together, getting close to their original levels of satisfaction only when their children leave home.[17] Despite what we read in the popular press, the only known symptom of 'empty nest syndrome' is increased smiling.[18] Interestingly, this pattern of satisfaction over the life cycle describes women (who are usually the primary caretakers of children) better than men.[19] Careful studies of how women feel as they go about their daily activities show that they are less happy when taking care of their children than when eating, exercising, shopping, napping, or watching television.[20] Indeed, looking after the kids appears to be only slightly more pleasant than doing housework.

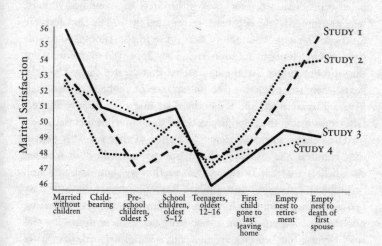

Fig. 23. As the four separate studies in this graph show, marital satisfaction decreases dramatically after the birth of the first child and increases only when the last child leaves home.

None of this should surprise us. Every parent knows that children are a lot of work – a lot of really *hard* work – and although parenting has many rewarding moments, the vast majority of its moments involve dull and selfless service to people who will take decades to become even begrudgingly grateful for what we are doing. If parenting is such difficult business, then why do we have such a rosy view of it? One reason is that we have been talking on the phone all day with society's stockholders – our mums and uncles and personal trainers – who have been transmitting to us an idea that they *believe* to be true but whose accuracy is not the cause of its successful transmission. 'Children bring happiness' is a super-replicator. The belief-transmission network of which we are a part cannot operate without a continuously replenished supply of people to do the transmitting, thus the belief that children are a source of happiness becomes a part of our cultural wisdom simply because the opposite belief unravels the fabric of any society that holds it. Indeed, people who believed that children bring misery and despair – and who thus stopped having them – would put their belief-transmission network out of business in around fifty years, hence terminating the belief that terminated them. The Shakers were a utopian farming community that arose in the 1800s and at one time numbered about six thousand. They approved of children, but they did not approve of the natural act that creates them. Over the years, their strict belief in the importance of celibacy caused their network to contract, and today there are just a few elderly Shakers left, transmitting their doomsday belief to no one but themselves.

The belief-transmission game is rigged so that we *must* believe that children and money bring happiness, regardless of whether such beliefs are true. This doesn't mean that we should all now quit our jobs and abandon our families. Rather, it means that while we *believe* we are raising children and earning paycheques to increase our share of happiness, we are actually doing these things for reasons beyond our ken. We are nodes in a social network that arises and falls by a logic of its own, which is why we continue to toil, continue to mate and continue to be surprised when we do not experience all the joy we so gullibly anticipated.

The Myth of Fingerprints

My friends tell me that I have a tendency to point out problems without offering solutions, but they never tell me what I should do about it. In one chapter after another, I've described the ways in which imagination fails to provide us with accurate previews of our emotional futures. I've claimed that when we imagine our futures we tend to fill in, leave out and take little account of how differently we will think about the future once we actually get there. I've claimed that neither personal experience nor cultural wisdom compensates for imagination's shortcomings. I've so thoroughly marinated you in the foibles, biases, errors and mistakes of the human mind that you may wonder how anyone ever manages to make toast without buttering their kneecaps. If so, you will be heartened to learn that there *is* a simple method by which anyone can make strikingly accurate predictions about how they will feel in the future. But you may be disheartened to learn that, by and large, no one wants to use it.

Why do we rely on our imaginations in the first place? Imagination is the poor man's wormhole. We can't do what we'd really *like* to do – namely, travel through time, pay a visit to our future selves and *see* how happy those selves are – and so we imagine the future instead of actually going there. But if we cannot travel in the dimension of time, we can travel in the dimensions of space, and the chances are pretty good that somewhere in those other three dimensions there is another human being who is actually *experiencing* the future event that we are merely thinking about. Surely we aren't the first people ever to consider a move to Cincinnati, a career in motel management, another helping of rhubarb pie or an extramarital affair, and for the most part, those who have already tried these things are more than willing to tell us about them. It is true that when people tell us about their past experiences ('That ice water wasn't really so cold' or 'I love taking care of my daughter'), memory's peccadilloes may render their testimony unreliable. But it is also true that when people tell us about their *current* experiences ('How am I feeling right now? I feel like pulling my arm out of this freezing bucket and sticking my teenager's head in it instead!'), they are providing us with the kind of report about their subjective state that is considered the gold standard of happiness measures. If you believe (as I do) that people can generally say how they are feeling at the moment they are asked, then one way to make predictions about our own emotional futures is to find someone who is having the experience we are contemplating and ask them how they feel. Instead of remembering our past experience in order to simulate our future experience, perhaps we should simply ask other people to introspect on their inner states. Perhaps we should give up on remembering and imagining entirely and use other people as *surrogates* for our future selves.

This idea sounds all too simple, and I suspect you have an objection to it that goes something like this: *Yes, other people are probably right now experiencing the very things I am merely contemplating, but I can't use other people's experiences as proxies for my own because those other people are not me. Every human being is as unique as his or her fingerprints, so it won't help me much to learn about how others feel in the situations that I'm facing. Unless these other people are my clones and have had all the same experiences I've had, their reactions and my reactions are bound to differ. I am a walking, talking idiosyncrasy, and thus I am better off basing my predictions on my somewhat fickle imagination than on the reports of people whose preferences, tastes and emotional proclivities are so radically different from my own.* If that's your objection, then it is a good one – so good that it will take two steps to dismantle it. First let me prove to you that the experience of a single randomly selected individual can sometimes provide a better basis for predicting your future experience than your own imagination can. And then let me show you why you – and I – find this so difficult to believe.

Finding the Solution

Imagination has three shortcomings, and if you didn't know that then you may be reading this book backwards. If you did know that, then you also know that imagination's first shortcoming is its tendency to fill in and leave out without telling us (which we explored in the section on *realism*). No one can imagine every feature and consequence of a future event, hence we must consider some and fail to consider others. The problem is that the features and consequences we fail to consider are often quite important. You may recall the study in which college students were asked to imagine how they would feel a few days after their school's football team played a game against its archrival.[21] The results showed that students overestimated the duration of the game's emotional impact because when they tried to imagine their future experience, they imagined their team winning ('The clock will hit zero, we'll storm the field, everyone will cheer . . .') but failed to imagine what they would be doing afterward ('And then I'll go home and study for my final exams'). Because the students were focused on the game, they failed to imagine how events that happened *after* the game would influence their happiness. So what *should* they have done instead?

They should have abandoned imagination altogether. Consider a study that put people in a similar predicament and then forced them to abandon their imaginations. In this study, a group of volunteers (reporters) first received a delicious prize – a gift certificate from a local ice cream parlour – and then performed a long, boring task in which they counted and recorded geometric shapes that appeared on a computer screen.[22] The reporters then reported how they felt. Next, a new group of volunteers was told that they would also receive a prize and do the same boring task. Some of these new volunteers (simulators) were told what the prize was and were asked to use their imaginations to predict their future feelings. Other volunteers (surrogators) were not told what the prize was but were instead shown the report of a randomly selected reporter. Not knowing what the prize was, they couldn't possibly use their imaginations to predict their future feelings. Instead, they had to rely on the reporter's report. Once all the volunteers had made their predictions, they received the prize, did the long, boring task and reported how they actually felt. As the leftmost bars in figure 24 show, simulators were not as happy as they thought they would be. Why? Because they failed to imagine how quickly the joy of receiving a gift certificate would fade when it was followed by a long, boring task. This is precisely the same mistake that the college-football fans made. But now look at the results for the surrogators. As you can see, they made extremely accurate predictions of their future happiness. These surrogators didn't know what kind of prize they would receive, but they did know that someone who had received that prize had been less than ecstatic at the conclusion of the boring task. So they shrugged and reasoned that they too would feel less than ecstatic at the conclusion of the boring task – and they were right!

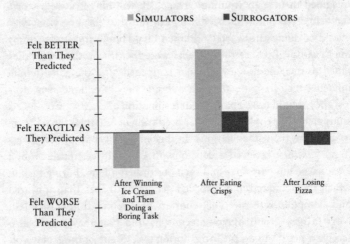

Fig. 24. Volunteers made much more accurate predictions of their future feelings when they learned how someone else had felt in the same situation (surrogators) than when they tried to imagine how they themselves would feel (simulators).

Imagination's second shortcoming is its tendency to project the present onto the future (which we explored in the section on *presentism*). When imagination paints a picture of the future, many of the details are necessarily missing, and imagination solves this problem by filling in the gaps with details that it borrows from the present. Anyone who has ever shopped on an empty stomach, vowed to quit smoking after stubbing out a cigarette or proposed marriage while on shore leave knows that how we feel now can erroneously influence how we *think* we'll feel later. As it turns out, surrogation can remedy this shortcoming too. In one study, volunteers (reporters) ate a few crisps and reported how much they enjoyed them.[23] Next, a new group of volunteers was fed pretzels, peanut-butter cheese crackers, tortilla chips, bread sticks and melba toast, which, as you might guess, left them thoroughly stuffed and with little desire for salty snack foods. These stuffed volunteers were then asked to predict how much they would enjoy eating a particular food the next day. Some of these stuffed volunteers (simulators) were told that the food they would eat the next day was crisps, and they were asked to use their imaginations to predict how they would feel after eating them. Other stuffed volunteers (surrogators) were not told what the next day's food would be but were instead shown the report of one randomly selected reporter. Because surrogators didn't know what the next day's food would be, they couldn't use their imaginations to predict their future enjoyment of it and thus they had to rely on the reporter's report. Once all the volunteers had made their predictions, they went away, returned the next day, ate some crisps and reported how much they enjoyed them. As the middle bars in figure 24 show, simulators enjoyed eating the crisps more than they thought they would. Why? Because when they made their predictions they had bellies full of pretzels and crackers. But surrogators – who were equally full when they made their predictions – relied on the report of someone without a full belly and hence made much more accurate predictions. It is important to note that the surrogators accurately predicted their future enjoyment of a food despite the fact that they didn't even know what the food was!

Imagination's third shortcoming is its failure to recognize that things will look different once they happen – in particular, that bad things will look a whole lot better (which we explored in the section on *rationalization*). When we imagine losing a job, for instance, we imagine the painful experience ('The boss will march into my office, shut the door behind him . . .') without also imagining how our psychological immune systems will transform its meaning ('I'll come to realize that this was an opportunity to quit retail sales and follow my true calling as a sculptor'). Can surrogation remedy this shortcoming? To find out, researchers arranged for some people to have an unpleasant experience. A group of volunteers (reporters) was told that the experimenter would flip a coin, and if it came up heads, the volunteer would receive a gift certificate to a local pizza restaurant. The coin was flipped and – *oh, so sorry* – it came up tails and the reporters received nothing.[24] The reporters then reported how they felt. Next, a new group of volunteers was told about the coin-flipping game and was asked to predict how they would feel if the coin came up tails and they didn't get the pizza gift certificate. Some of these volunteers (simulators) were told the precise monetary value of the gift certificate, and others (surrogators) were instead shown the report of one randomly selected reporter. Once the volunteers had made their predictions, the coin was flipped and – *oh, so sorry* – came up tails. The volunteers then reported how they felt. As the rightmost bars in figure 24 show, simulators felt better than they predicted they'd feel if they lost the coin flip. Why? Because simulators did not realize how quickly and easily they would rationalize the loss ("Pizza is too fattening, and besides, I don't like that restaurant anyway"). But surrogators – who had nothing to go on except the report of another randomly selected individual – assumed that they wouldn't feel too bad after losing the prize, hence made more accurate predictions.

Rejecting the Solution

This trio of studies suggests that when people are deprived of the information that imagination requires and are thus *forced* to use others as surrogates, they make remarkably accurate predictions about their future feelings, which suggests that the best way to predict our feelings tomorrow is to see how others are feeling today.[25] Given the impressive power of this simple technique, we should expect people to go out of their way to use it. But they don't. When an entirely new group of volunteers was told about the three situations I just described – winning a prize, eating a mystery food or failing to receive a gift certificate – and was then asked whether they would prefer to make predictions about their future feelings based on *(a)* information about the prize, the food and the certificate; or *(b)* information about how a randomly selected individual felt after winning them, eating them or losing them, virtually every volunteer chose the former. If you hadn't seen the results of these studies, you'd probably have done the same. If I offered to pay for your dinner at a restaurant if you could accurately predict how much you were going to enjoy it, would you want to see the restaurant's menu or some randomly selected diner's review? If you are like most people, you would prefer to see the menu, and if you are like most people, you would end up buying your own dinner. Why?

Because if you are like most people, then like most people, you don't know you're like most people. Science has given us a lot of facts about the average person, and one of the most reliable of these facts is that the average person doesn't see herself as average. Most students see themselves as more intelligent than the average student,[26] most business managers see themselves as more competent than the average business manager,[27] and most football players see themselves as having better 'football sense' than their teammates.[28] Ninety per cent of motorists consider themselves to be safer-than-average drivers,[29] and 94 per cent of college professors consider themselves to be better-than-average teachers.[30] Ironically, the bias toward seeing ourselves as better than average causes us to see ourselves as less biased than average too.[31] As one research team concluded, 'Most of us appear to believe that we are more athletic, intelligent, organized, ethical, logical, interesting, fair-minded, and healthy – not to mention more attractive – than the average person.'[32]

This tendency to think of ourselves as better than others is not necessarily a manifestation of our unfettered narcissism but may instead be an instance of a more general tendency to think of ourselves as *different* from others – often for better but sometimes for worse. When people are asked about generosity, they claim to perform a greater number of generous acts than others do; but when they are asked about selfishness, they claim to perform a greater number of selfish acts than others do.[33] When people are asked about their ability to perform an easy task, such as driving a car or riding a bike, they rate themselves as better than others; but when they are asked about their ability to perform a difficult task, such as juggling or playing chess, they rate themselves as worse than others.[34] We don't always see ourselves as *superior*, but we almost always see ourselves as *unique*. Even when we do precisely what others do, we tend to think that we're doing it for unique reasons. For instance, we tend to attribute other people's choices to features of the chooser ('Phil picked this class because he's one of those literary types'), but we tend to attribute our own choices to features of the options ('But I picked it because it was easier than economics').[35] We recognize that our decisions are influenced by social norms ('I was too embarrassed to raise my hand in class even though I was terribly confused'), but fail to recognize that others' decisions were similarly influenced ('No one else raised a hand because no one else was as confused as I was').[36] We know that our choices sometimes reflect our aversions ('I voted for Kerry because I couldn't stand Bush'), but we assume that other people's choices reflect their appetites ('If Rebecca voted for Kerry, then she must have liked him').[37] The list of differences is long but the conclusion to be drawn from it is short: the self considers itself to be a very special person.[38]

What makes us think we're so damned special? Three things, at least. First, even if we aren't special, the way we know ourselves is. We are the only people in the world whom we can know from the inside. We *experience* our own thoughts and feelings but must *infer* that other people are experiencing theirs. We all trust that behind those eyes and inside those skulls, our friends and neighbours are having subjective experiences very much like our own, but that trust is an article of faith and not the palpable, self-evident truth that our own subjective experiences constitute. There is a difference between making love and reading about it, and it is the same difference that distinguishes our knowledge of our own mental lives from our knowledge of everyone else's. Because we know ourselves and others by such different means, we gather very different kinds and amounts of information. In every waking moment we monitor the steady stream of thoughts and feelings that runs through our heads, but we only monitor other people's words and deeds, and only when they are in our company. One reason why we seem so special, then, is that we learn about ourselves in such a special way.

The second reason is that we *enjoy* thinking of ourselves as special. Most of us want to fit in well with our peers, but we don't want to fit in too well.[39] We prize our unique identities, and research shows that when people are made to feel too similar to others, their moods quickly sour and they try to distance and distinguish themselves in a variety of ways.[40] If you've ever shown up at a party and found someone else wearing exactly the same dress or tie that you were wearing, then you know how unsettling it is to share the room with an unwanted twin whose presence temporarily diminishes your sense of individuality. Because we *value* our uniqueness, it isn't surprising that we tend to overestimate it.

The third reason why we tend to overestimate our uniqueness is that we tend to overestimate everyone's uniqueness – that is, we tend to think of people as more different from one another than they actually are. Let's face it: all people are similar in some ways and different in others. The psychologists, biologists, economists and sociologists who are searching for universal laws of human behavior naturally care about the similarities, but the rest of us care mainly about the differences. Social life involves selecting particular individuals to be our sexual partners, business partners, bowling partners, and more. That task requires that we focus on the things that distinguish one person from another and not on the things that all people share, which is why personal ads are much more likely to mention the advertiser's love of ballet than his love of oxygen. A penchant for respiration explains a great deal about human behaviour – for example, why people live on land, become ill at high altitudes, have lungs, resist suffocation, love trees, and so on. It surely explains more than does a person's penchant for ballet. But it does nothing to distinguish one person from another, and thus for ordinary people who are in the ordinary business of selecting others for commerce, conversation or copulation, the penchant for air is stunningly irrelevant. Individual similarities are vast, but we don't care much about them because they don't help us do what we are here on earth to do, namely, distinguish Jack from Jill and Jill from Jennifer. As such, these individual similarities are an inconspicuous backdrop against which a small number of relatively minor individual differences stand out in bold relief.

Because we spend so much time searching for, attending to, thinking about and remembering these differences, we tend to overestimate their magnitude and frequency, and thus end up thinking of people as more varied than they actually are. If you spent all day sorting grapes into different shapes, colours and kinds, you'd become one of those annoying grapeophiles who talks endlessly about the nuances of flavour and the permutations of texture. You'd come to think of grapes as infinitely varied, and you'd forget that almost all of the really *important* information about a grape can be deduced from the simple fact of its grapehood. Our belief in the variability of others and in the uniqueness of the self is especially powerful when it comes to emotion.[41] Because we can *feel* our own emotions but must *infer* the emotions of others by watching their faces and listening to their voices, we often have the impression that others don't experience the same intensity of emotion that we do, which is why we expect others to recognize our feelings even when we can't recognize theirs.[42] This sense of emotional uniqueness starts early. When kindergarteners are asked how they and others would feel in a variety of situations, they expect to experience unique emotions ('Billy would be sad but I wouldn't') and they provide unique reasons for experiencing them ('I'd tell myself that the hamster was in heaven, but Billy would just cry').[43] When adults make these same kinds of prediction, they do just the same thing.[44]

Our mythical belief in the variability and uniqueness of individuals is the main reason why we refuse to use others as surrogates. After all, surrogation is only useful when we can count on a surrogate to react to an event roughly as we would, and if we believe that people's emotional reactions are more varied than they actually are, then surrogation will seem less useful to us than it actually is. The irony, of course, is that surrogation is a cheap and effective way to predict one's future emotions, but because we don't realize just how similar we all are, we reject this reliable method and rely instead on our imaginations, as flawed and fallible as they may be.

Onward

Despite its watery connotation, the word *hogwash* refers to the feeding – and not to the bathing – of pigs. Hogwash is something that pigs eat, that pigs like and that pigs need. Farmers provide pigs with hogwash because without it, pigs get grumpy. The word *hog-*

wash also refers to the falsehoods people tell one another. Like the hogwash that farmers feed their pigs, the hogwash that our friends and teachers and parents feed us is meant to make us happy; but unlike hogwash of the porcine variety, human hogwash does not always achieve its end. As we have seen, ideas can flourish if they preserve the social systems that allow them to be transmitted. Because individuals don't usually feel that it is their personal duty to preserve social systems, these ideas must disguise themselves as prescriptions for individual happiness. We might expect that after spending some time in the world, our experiences would debunk these ideas, but it doesn't always work that way. To learn from our experience we must remember it, and for a variety of reasons, memory is a faithless friend. Practice and coaching get us out of our nappies and into our britches, but they are not enough to get us out of our presents and into our futures. What's so ironic about this predicament is that the information we need to make accurate predictions of our emotional futures is right under our noses, but we don't seem to recognize its aroma. It doesn't always make sense to heed what people tell us when they communicate their beliefs about happiness, but it does make sense to observe how happy they are in different circumstances. Alas, we think of ourselves as unique entities – minds unlike any others – and thus we often reject the lessons that the emotional experience of others has to teach us.

AFTERWORD

My mind presageth happy gain and conquest.

Shakespeare, *King Henry VI, Part III*

MOST OF US MAKE at least three important decisions in our lives: where to live, what to do and with whom to do it. We choose our towns and our neighbourhoods, we choose our jobs and our hobbies, we choose our spouses and our friends. Making these decisions is such a natural part of adulthood that it is easy to forget that we are among the first human beings to make them. For most of recorded history, people lived where they were born, did what their parents had done and associated with those who were doing the same. Millers milled, Smiths smithed, and little Smiths and little Millers married whom and when they were told. Social structures (such as religions and castes) and physical structures (such as mountains and oceans) were the great dictators that determined how, where and with whom people would spend their lives, which left most people with little to decide for themselves. But the agricultural, industrial and technological revolutions changed all that, and the resulting explosion of personal liberty has created a bewildering array of options, alternatives, choices and decisions that our ancestors never faced. For the very first time, our happiness is in our hands.

How are we to make these choices? In 1738, a Dutch polymath named Daniel Bernoulli claimed he had the answer. He suggested that the wisdom of any decision could be calculated by multiplying the *probability* that the decision will give us what we want by the *utility* of getting what we want. By *utility*, Bernoulli meant some-

thing like *goodness* or *pleasure*.[1] The first part of Bernoulli's pre-
scription is fairly easy to follow because in most circumstances we
can roughly estimate the odds that our choices will get us where we
want to be. How likely is it that you'll be promoted to general manager if you take the job at IBM? How likely is it that you'll spend
your weekends at the beach if you move to St. Petersburg? How
likely is it that you'll have to sell your motorcycle if you marry
Eloise? Calculating such odds is relatively straightforward stuff,
which is why insurance companies get rich by doing little more than
estimating the likelihood that your house will burn down, your car
will be stolen and your life will end early. With a little detective
work, a pencil and a good eraser, we can usually estimate – at least
roughly – the probability that a choice will give us what we desire.

The problem is that we cannot easily estimate how we'll feel
when we get it. Bernoulli's brilliance lay not in his mathematics but
in his psychology – in his realization that what we objectively get
(*wealth*) is not the same as what we subjectively *experience* when we
get it (*utility*). Wealth may be measured by counting dollars, but utility must be measured by counting how much goodness those dollars
buy.[2] Wealth doesn't matter; utility does. We don't care about money
or promotions or beach holidays per se; we care about the goodness
or pleasure that these forms of wealth may (or may not) induce.
Wise choices are those that maximize our pleasure, not our dollars,
and if we are to have any hope of choosing wisely, then we must
correctly anticipate how much pleasure those dollars will buy us.
Bernoulli knew that it was much easier to predict how much wealth
a choice might produce than how much utility a choice might produce, so he devised a simple conversion formula that he hoped
would allow anyone to translate estimates of the former into estimates of the latter. He suggested that each successive dollar provides
a little less pleasure than the one before it, and that a person can
therefore calculate the pleasure they will derive from a dollar simply
by correcting for the number of dollars they already have.

> The determination of the *value* of an item must not be based
> on its *price*, but rather on the *utility* it yields. The price of the
> item is dependent only on the thing itself and is equal for

everyone; the utility, however, is dependent on the particular circumstances of the person making the estimate. Thus there is no doubt that a gain of one thousand ducats is more significant to a pauper than to a rich man though both gain the same amount.[3]

Bernoulli correctly realized that people are sensitive to relative rather than absolute magnitudes, and his formula was meant to take this basic psychological truth into account. But he also knew that translating wealth into utility was not as simple as he'd made it out to be, and that there were other psychological truths that his formula ignored.

Although a poor man generally obtains more utility than does a rich man from an equal gain, it is nevertheless conceivable, for example, that a rich prisoner who possesses two thousand ducats but needs two thousand ducats more to repurchase his freedom, will place a higher value on a gain of two thousand ducats than does another man who has less money than he. Though innumerable examples of this kind may be constructed, they represent exceedingly rare exceptions.[4]

It was a good try. Bernoulli was right in thinking that the hundredth dollar (or kiss or doughnut or romp in the meadow) generally does not make us as happy as the first one did, but he was wrong in thinking that this was the *only* thing that distinguished wealth from utility and hence the only thing one must correct for when predicting utility from wealth. As it turns out, the 'innumerable exceptions' that Bernoulli swept under the rug are *not* exceedingly rare. There are *many* things other than the size of a person's bank account that influence how much utility they derive from the next dollar. For instance, people often value things more after they own them than before, they often value things more when they are imminent than distant, they are often hurt more by small losses than by large ones, they often imagine that the pain of losing something is greater than the pleasure of getting it, and so on – and on and on and on. The myriad phenomena with which this book has been concerned are just some of the not-so-rare exceptions that make Bernoulli's principle a beautiful, useless abstraction. Yes, we *should* make choices by multiplying probabilities and utilities, but how can we possibly do this if we can't estimate those utilities beforehand? The same objective circumstances give rise to a remarkably wide variety of subjective experiences, and thus it is very difficult to predict our subjective experiences from foreknowledge of our objective circumstances. The sad fact is that converting wealth to utility – that is, predicting how we will feel from knowledge of what we will get – isn't very much like converting metres to yards or German to Japanese. The simple, lawful relationships that bind numbers to numbers and words to words do not bind objective events to emotional experiences.

So what's a chooser to do? Without a formula for predicting utility, we tend to do what only our species does: imagine. Our brains have a unique structure that allows us to mentally transport ourselves into future circumstances and then ask ourselves how it feels to be there. Rather than calculating utilities with mathematical precision, we simply step into tomorrow's shoes and see how well they fit. Our ability to project ourselves forward in time and experience events before they happen enables us to learn from mistakes without making them and to evaluate actions without taking them. If nature has given us a greater gift, no one has named it. And yet, as impressive as it is, our ability to simulate future selves and future circumstances is by no means perfect. When we imagine future circumstances, we fill in details that won't really come to pass and leave out details that will. When we imagine future feelings, we find it impossible to ignore what we are feeling now and impossible to recognize how we will think about the things that happen later. Daniel Bernoulli dreamed of a world in which a simple formula would allow us all to determine our futures with perspicacity and foresight. But foresight is a fragile talent that often leaves us squinting, straining to see what it would be like to have this, go there or do that. There is no simple formula for finding happiness. But if our great big brains do not allow us to go surefootedly into our futures, they at least allow us to understand what makes us stumble.

NOTES

Foreword

1. The notes in this book contain references to the scientific research that supports the claims I make in the text. Occasionally they contain some extra information that may be of interest but that is not essential to the argument. If you don't care about sources, aren't interested in nonessentials and are annoyed by books that make you flip back and forth all the time, then be assured that the only important note in the book is this one.

Chapter 1: Journey to Elsewhen

1. W. A. Roberts, 'Are Animals Stuck in Time?', *Psychological Bulletin* 128: 473–89 (2002).

2. D. Dennett, *Kinds of Minds* (New York: Basic Books, 1996).

3. M. M. Haith, 'The Development of Future Thinking as Essential for the Emergence of Skill in Planning', in *The Developmental Psychology of Planning: Why, How, and When Do We Plan?*, ed. S. L. Friedman and E. K. Scholnick (Mahwah, NJ: Lawrence Erlbaum, 1997), 25–42.

4. E. Bates, J. Elman and P. Li, 'Language In, On, and About Time', in *The Development of Future Oriented Processes*, ed. M. M. Haith et al. (Chicago: University of Chicago Press, 1994).

5. B. M. Hood et al., 'Gravity Biases in a Nonhuman Primate?', *Developmental Science* 2: 35–41 (1999). See also D. A. Washburn and D. M. Rumbaugh, 'Comparative Assessment of Psychomotor Performance: Target Prediction by Humans and Macaques *(Macaca mulatta)*', *Journal of Experimental Psychology: General* 121: 305–12 (1992).

6. L. M. Oakes and L. B. Cohen, 'Infant Perception of a Causal Event', *Cognitive Development* 5: 193–207 (1990). See also N. Wentworth and M. M. Haith, 'Event-Specific Expectations of 2- and 3-Month-Old Infants', *Developmental Psychology* 28: 842–50 (1992).

7. C. M. Atance and D. K. O'Neill, 'Planning in 3-Year-Olds: A Reflection of the Future Self?', in *The Self in Time: Developmental Perspectives*, ed. C. Moore and K. Lemmon (Mahwah, NJ: Lawrence Erlbaum, 2001); and J. B. Benson, 'The Development of Planning: It's About Time', in Friedman and Scholnick, *Developmental Psychology of Planning*.

8. Although children begin to talk about the future at about two years, they don't seem to have a full understanding of it until about age four. See D. J. Povinelli and B. B. Simon, 'Young Children's Understanding of Briefly Ver-

sus Extremely Delayed Images of the Self: Emergence of the Autobiographical Stance', *Developmental Psychology* 34: 188–94 (1998); and K. Nelson, 'Finding One's Self in Time', in *The Self Across Psychology: Self-Recognition, Self-Awareness, and the Self Concept*, ed. J. G. Snodgrass and R. L. Thompson (New York: New York Academy of Sciences, 1997), 103–16.

9. C. A. Banyas, 'Evolution and Phylogenetic History of the Frontal Lobes', in *The Human Frontal Lobes*, ed. B. L. Miller and J. L. Cummings (New York: Guilford Press, 1999), 83–106.

10. Phineas apparently took the rod with him wherever he went for the rest of his life and would probably be pleased that both it and his skull ended up on permanent display at Harvard's Warren Anthropological Museum.

11. Modern authors cite the Gage case as evidence for the *importance* of the frontal lobe, but this is not the way people thought about the incident when it happened. See M. B. Macmillan, 'A Wonderful Journey Through Skull and Brains: The Travels of Mr. Gage's Tamping Iron', *Brain and Cognition* 5: 67–107 (1986).

12. M. B. Macmillan, 'Phineas Gage's Contribution to Brain Surgery', *Journal of the History of the Neurosciences* 5: 56–77 (1996).

13. S. M. Weingarten, 'Psychosurgery', in Miller and Cummings, *Human Frontal Lobes*, 446–60.

14. D. R. Weinberger et al., 'Neural Mechanisms of Future-Oriented Processes', in Haith et al., *Development of Future Oriented Processes*, 221–42.

15. J. M. Fuster, *The Prefrontal Cortex: Anatomy, Physiology, and Neuropsychology of the Frontal Lobe* (New York: Lippincott-Raven, 1997), 160–61.

16. A. K. MacLeod and M. L. Cropley, 'Anxiety, Depression, and the Anticipation of Future Positive and Negative Experiences', *Journal of Abnormal Psychology* 105: 286–89 (1996).

17. M. A. Wheeler, D. T. Stuss and E. Tulving, 'Toward a General Theory of Episodic Memory: The Frontal Lobes and Autonoetic Consciousness', *Psychological Bulletin* 121: 331–54 (1997).

18. F. T. Melges, 'Identity and Temporal Perspective', in *Cognitive Models of Psychological Time*, ed. R. A. Block (Hillsdale, NJ: Lawrence Erlbaum, 1990), 255–66.

19. P. Faglioni, 'The Frontal Lobes', in *The Handbook of Clinical and Experimental Neuropsychology*, ed. G. Denes and L. Pizzamiglio (East Sussex, UK: Psychology Press, 1999), 525–69.

20. J. M. Fuster, 'Cognitive Functions of the Frontal Lobes', in Miller and Cummings, *Human Frontal Lobes*, 187–95.

21. E. Tulving, 'Memory and Consciousness', *Canadian Psychology* 26: 1–12 (1985). The same case is described extensively under the pseudonym 'K.C.' in E. Tulving et al., 'Priming of Semantic Autobiographical Knowledge: A Case Study of Retrograde Amnesia', *Brain and Cognition* 8: 3–20 (1988).

22. Tulving, 'Memory and Consciousness'.

23. R. Dass, *Be Here Now* (New York: Crown, 1971).

24. L. A. Jason et al., 'Time Orientation: Past, Present, and Future Perceptions', *Psychological Reports* 64: 1199–1205 (1989).

25. E. Klinger and W. M. Cox, 'Dimensions of Thought Flow in Everyday Life', *Imagination, Cognition, and Personality* 72: 105–28 (1987–88); and E. Klinger, 'On Living Tomorrow Today: The Quality of Inner Life as a Function of Goal Expectations', in *Psychology of Future Orientation*, ed. Z. Zaleski (Lublin, Poland: Towarzystwo Naukowe KUL, 1994), 97–106.

26. J. L. Singer, *Daydreaming and Fantasy* (Oxford: Oxford University Press, 1981); E. Klinger, *Daydreaming: Using Waking Fantasy and Imagery for Self-Knowledge and Creativity* (Los Angeles: Tarcher, 1990); G. Oettingen, *Psychologie des Zukunftdenkens* [On the Psychology of Future Thought] (Göttingen, Germany: Hogrefe, 1997).

27. G. F. Loewenstein and D. Prelec, 'Preferences for Sequences of Outcomes', *Psychological Review* 100: 91–108 (1993). See also G. Loewenstein, 'Anticipation and the Valuation of Delayed Consumption', *Economy Journal* 97: 666–84 (1987); and J. Elster and G. F. Loewenstein, 'Utility from Memory and Anticipation', in *Choice Over Time*, ed. G. F. Loewenstein and J. Elster (New York: Russell Sage Foundation, 1992), 213–34.

28. G. Oettingen and D. Mayer, 'The Motivating Function of Thinking About the Future: Expectations Versus Fantasies', *Journal of Personality and Social Psychology* 83: 1198–1212 (2002).

29. A. Tversky and D. Kahneman, 'Availability: A Heuristic for Judgment Frequency and Probability', *Cognitive Psychology* 5: 207–32 (1973).

30. N. Weinstein, 'Unrealistic Optimism About Future Life Events', *Journal of Personality and Social Psychology* 39: 806–20 (1980).

31. P. Brickman, D. Coates and R. J. Janoff-Bulman, 'Lottery Winners and Accident Victims: Is Happiness Relative?', *Journal of Personality and Social Psychology* 36: 917–27 (1978).

32. E. C. Chang, K. Asakawa and L. J. Sanna, 'Cultural Variations in Optimistic and Pessimistic Bias: Do Easterners Really Expect the Worst and Westerners Really Expect the Best When Predicting Future Life Events?', *Journal of Personality and Social Psychology* 81: 476–91 (2001).

33. J. M. Burger and M. L. Palmer, 'Changes in and Generalization of Unrealistic Optimism Following Experiences with Stressful Events: Reactions to the 1989 California Earthquake', *Personality and Social Psychology Bulletin* 18: 39–43 (1992).

34. H. E. Stiegelis et al., 'Cognitive Adaptation: A Comparison of Cancer Patients and Healthy References', *British Journal of Health Psychology* 8: 303–18 (2003).

35. A. Arntz, M. Van Eck and P. J. de Jong, 'Unpredictable Sudden Increases in Intensity of Pain and Acquired Fear', *Journal of Psychophysiology* 6: 54–64 (1992).

36. Speaking of electric shock, this is probably a good time to mention that psychological experiments such as these are always performed according to the strict ethical guidelines of the American Psychological Association and must be approved by university committees before they are implemented. Those who participate do so voluntarily, are always fully informed of any risks the study may pose to their health or happiness and are given the opportunity to withdraw at

any time without fear of penalty. If people are given any false information in the course of an experiment, they are told the truth when the experiment is over. In short, we're really *very* nice people.

37. M. Miceli and C. Castelfranchi, 'The Mind and the Future: The (Negative) Power of Expectations', *Theory and Psychology* 12: 335–66 (2002).

38. J. N. Norem, 'Pessimism: Accentuating the Positive Possibilities', in *Virtue, Vice, and Personality: The Complexity of Behavior*, ed. E. C. Chang and L. J. Sanna (Washington, DC: American Psychological Association, 2003), 91–104; J. K. Norem and N. Cantor, 'Defensive Pessimism: Harnessing Anxiety as Motivation', *Journal of Personality and Social Psychology* 51: 1208–17 (1986).

39. A. Bandura, 'Self-Efficacy: Toward a Unifying Theory of Behavioral Change', *Psychological Review* 84: 191–215 (1977); and A. Bandura, 'Self-Efficacy: Mechanism in Human Agency', *American Psychologist* 37: 122–47 (1982).

40. M. E. P. Seligman, *Helplessness: On Depression, Development, and Death* (San Francisco: Freeman, 1975).

41. E. Langer and J. Rodin, 'The Effect of Choice and Enhanced Personal Responsibility for the Aged: A Field Experiment in an Institutional Setting', *Journal of Personality and Social Psychology* 34: 191–98 (1976); and J. Rodin and E. J. Langer, 'Long-Term Effects of a Control-Relevant Intervention with the Institutional Aged', *Journal of Personality and Social Psychology* 35: 897–902 (1977).

42. R. Schulz and B. H. Hanusa, 'Long-Term Effects of Control and Predictability-Enhancing Interventions: Findings and Ethical Issues', *Journal of Personality and Social Psychology* 36: 1202–12 (1978).

43. E. J. Langer, 'The Illusion of Control', *Journal of Personality and Social Psychology* 32: 311–28 (1975).

44. Ibid.

45. D. S. Dunn and T. D. Wilson, 'When the Stakes Are High: A Limit to the Illusion of Control Effect', *Social Cognition* 8: 305–23 (1991).

46. L. H. Strickland, R. J. Lewicki and A. M. Katz, 'Temporal Orientation and Perceived Control as Determinants of Risk Taking', *Journal of Experimental Social Psychology* 2: 143–51 (1966).

47. Dunn and Wilson, 'When the Stakes Are High'.

48. S. Gollin et al., 'The Illusion of Control Among Depressed Patients', *Journal of Abnormal Psychology* 88: 454–57 (1979).

49. L. B. Alloy and L. Y. Abramson, 'Judgment of Contingency in Depressed and Nondepressed Students: Sadder but Wiser?', *Journal of Experimental Psychology: General* 108: 441–85 (1979). For a contrary view, see D. Dunning and A. L. Story, 'Depression, Realism and the Overconfidence Effect: Are the Sadder Wiser When Predicting Future Actions and Events?', *Journal of Personality and Social Psychology* 61: 521–32 (1991); and R. M. Msetfi et al., 'Depressive Realism and Outcome Density Bias in Contingency Judgments: The Effect of the Context and Intertrial Interval', *Journal of Experimental Psychology: General* 134: 10–22 (2005).

50. S. E. Taylor and J. D. Brown, 'Illusion and Well-Being: A Social-Psychological Perspective on Mental Health', *Psychological Bulletin* 103: 193–210 (1988).

Chapter 2: The View from in Here

1. N. L. Segal, *Entwined Lives: Twins and What They Tell Us About Human Behavior* (New York: Dutton, 1999).

2. N. Angier, 'Joined for Life, and Living Life to the Full', *New York Times*, 23 December 1997, F1.

3. A. D. Dreger, 'The Limits of Individuality: Ritual and Sacrifice in the Lives and Medical Treatment of Conjoined Twins', *Studies in History and Philosophy of Biological and Biomedical Sciences* 29: 1–29 (1998).

4. Ibid. Since this paper was published, at least one pair of adult conjoined twins sought separation and died during surgery. 'A Lost Surgical Gamble', *New York Times*, 9 July 2003, 20.

5. J. R. Searle, *Mind, Language, and Society: Philosophy in the Real World* (New York: Basic Books, 1998).

6. Subjective states can be defined only in terms of their objective antecedents or other subjective states, but the same is true for physical objects. If we were not allowed to define a physical object (Marshmallow Fluf) in terms of the subjective states it brought about ('It's soft, gooey, and sweet') or in terms of any other physical object ('It's made from corn syrup, sugar syrup, vanilla flavouring and egg whites'), we could not define it. All definitions are achieved by comparing the thing we wish to define with things that inhabit the same ontological category (e.g., physical things to physical things) or by mapping them onto things in a different ontological category (e.g., physical things to subjective states). No one has yet discovered a third way.

7. R. D. Lane et al., 'Neuroanatomical Correlates of Pleasant and Unpleasant Emotion', *Neuropsychologia* 35: 1437–44 (1997).

8. C. Osgood, G. J. Suci and P. H. Tannenbaum, *The Measurement of Meaning* (Urbana: University of Illinois Press, 1957). The typical finding is that words differ on three dimensions: evaluation (good or bad); activity (active or passive); and potency (strong or weak). So psychologists talk about a word's E-ness, A-ness and P-ness. Say these terms aloud and then tell me that scientists have no sense of humour.

9. T. Nagel, 'What Is It Like to Be a Bat?', *Philosophical Review* 83: 435–50 (1974).

10. See A. Pope, *Essay on Man, Epistle 4* (1744), in *The Complete Poetical Works of Alexander Pope*, ed. H. W. Boynton (New York: Houghton Mifflin, 1903).

11. S. Freud, *Civilization and Its Discontents*, vol. 1 of *The Standard Edition of the Complete Psychological Works of Sigmund Freud* (1930; London: Hogarth Press and Institute of Psychoanalysis, 1953), 75–76.

12. B. Pascal, 'Pensees', in *Pensees*, ed. W. F. Trotter (1660; New York: Dutton, 1908).

13. R. Nozick, *The Examined Life* (New York: Simon & Schuster, 1989), 102.

14. J. S. Mill, 'Utilitarianism' (1863), in *On Liberty, the Subjection of Women and Utilitarianism,* in *The Basic Writings of John Stuart Mill,* ed. D. E. Miller (New York: Modern Library, 2002).

15. R. Nozick, *Anarchy, State, and Utopia* (New York: Basic Books, 1974).

16. Nozick, *The Examined Life*, p. 111. Nozick's 'happiness machine' problem is popular among academics, who generally fail to consider three things. First, who *says* that no one would want to be hooked up? The world is full of people who want happiness and don't care one bit about whether it is 'well deserved'. Second, those who claim that they would not agree to be hooked up may already be hooked up. After all, the deal is that you forget your previous decision. Third, no one can *really* answer this question because it requires them to imagine a future state in which they do not know the very thing they are currently contemplating. See E. B. Royzman, K. W. Cassidy and J. Baron, "I Know, You Know": Epistemic Egocentrism in Children and Adults', *Review of General Psychology* 7: 38–65 (2003).

17. D. M. MacMahon, 'From the Happiness of Virtue to the Virtue of Happiness: 400 BC–AD 1780', *Daedalus: Journal of the American Academy of Arts and Sciences* 133: 5–17 (2004).

18. Ibid.

19. For some discussions of the distinction between moral and emotional happiness, all of which take a position contrary to mine, see D. W. Hudson, *Happiness and the Limits of Satisfaction* (London: Rowman & Littlefield, 1996); M. Kingwell, *Better Living: In Pursuit of Happiness from Plato to Prozac* (Toronto: Viking, 1998); and E. Telfer, *Happiness* (New York: St. Martin's Press, 1980).

20. N. Block, 'Begging the Question Against Phenomenal Consciousness', *Behavioral and Brain Sciences* 15: 205–6 (1992).

21. J. W. Schooler and T. Y. Engstler-Schooler, 'Verbal Overshadowing of Visual Memories: Some Things Are Better Left Unsaid', *Cognitive Psychology* 22: 36–71 (1990.)

22. G. W. McConkie and D. Zola, 'Is Visual Information Integrated in Successive Fixations in Reading?', *Perception and Psychophysics* 25: 221–24 (1979).

23. D. J. Simons and D. T. Levin, 'Change Blindness', *Trends in Cognitive Sciences* 1: 261–67 (1997).

24. M. R. Beck, B. L. Angelone and D. T. Levin, 'Knowledge About the Probability of Change Affects Change Detection Performance', *Journal of Experimental Psychology: Human Perception and Performance* 30: 778–91 (2004).

25. D. J. Simons and D. T. Levin, 'Failure to Detect Changes to People in a Real-World Interaction', *Psychonomic Bulletin and Review* 5: 644–49 (1998).

26. R. A. Rensink, J. K. O'Regan and J. J. Clark, 'To See or Not to See: The Need for Attention to Perceive Changes in Scenes', *Psychological Science* 8: 368–73 (1997).

27. 'Hats Off to the Amazing Hondo', www.hondomagic.com/html/a_little_magic.htm.

28. B. Fischoff, 'Perceived Informativeness of Facts', *Journal of Experimental Psychology: Human Perception and Performance* 3: 349–58 (1977).

29. A. Parducci, *Happiness, Pleasure, and Judgment: The Contextual Theory and Its Applications* (Mahwah, NJ: Lawrence Erlbaum, 1995).

30. E. Shackleton, *South* (1959; New York: Carroll & Graf, 1998), 192.

Chapter 3: Outside Looking In

1. J. LeDoux, *The Emotional Brain: The Mysterious Underpinnings of Emotional Life* (New York: Simon & Schuster, 1996).

2. R. B. Zajonc, 'Feeling and Thinking: Preferences Need No Inferences', *American Psychologist* 35: 151–75 (1980); R. B. Zajonc, 'On the Primacy of Affect', *American Psychologist* 39: 117–23 (1984); and 'Emotions', in *The Handbook of Social Psychology*, ed. D. T. Gilbert, S. T. Fiske and G. Lindzey, 4th edn., vol. 1 (New York: McGraw-Hill, 1998), 591–632.

3. S. Schachter and J. Singer, 'Cognitive, Social and Physiological Determinants of Emotional State', *Psychological Review* 69: 379–99 (1962).

4. D. G. Dutton and A. P. Aron, 'Some Evidence for Heightened Sexual Attraction Under Conditions of High Anxiety', *Journal of Personality and Social Psychology* 30: 510–17 (1974).

5. It is also interesting to note that the mere act of identifying an emotion can sometimes eliminate it. See A. R. Hariri, S. Y. Bookheimer and J. C. Mazziotta, 'Modulating Emotional Response: Effects of a Neocortical Network on the Limbic System', *NeuroReport* 11: 43–48 (2000); and M. D. Lieberman et al., 'Two Captains, One Ship: A Social Cognitive Neuroscience Approach to Disrupting Automatic Affective Processes' (unpublished manuscript, UCLA, 2003).

6. G. Greene, *The End of the Affair* (New York: Viking Press, 1951), 29.

7. R. A. Dienstbier and P. C. Munter, 'Cheating as a Function of the Labeling of Natural Arousal', *Journal of Personality and Social Psychology* 17: 208–13 (1971).

8. M. P. Zanna and J. Cooper, 'Dissonance and the Pill: An Attribution Approach to Studying the Arousal Properties of Dissonance', *Journal of Personality and Social Psychology* 29: 703–9 (1974).

9. D. C. Dennett, *Brainstorms: Philosophical Essays on Mind and Psychology* (Cambridge, Mass.: Bradford/MIT Press, 1981), 218.

10. J. W. Schooler, 'Re-representing Consciousness: Dissociations Between Consciousness and Meta-Consciousness', *Trends in Cognitive Science* 6: 339–44 (2002).

11. L. Weiskrantz, *Blindsight* (Oxford: Oxford University Press, 1986).

12. A. Cowey and P. Stoerig, 'The Neurobiology of Blindsight', *Trends in Neuroscience* 14: 140–45 (1991).

13. E. J. Vanman, M. E. Dawson and P. A. Brennan, 'Affective Reactions in the Blink of an Eye: Individual Differences in Subjective Experience and Physiological Responses to Emotional Stimuli', *Personality and Social Psychology Bulletin* 24: 994–1005 (1998).

14. R. D. Lane et al., 'Is Alexithymia the Emotional Equivalent of Blindsight?', *Biological Psychiatry* 42: 834–44 (1997).

15. The nineteenth-century economist Francis Edgeworth referred to this device as a *hedonimeter*. See F. Y. Edgeworth, *Mathematical Psychics: An Essay on the Application of Mathematics to the Moral Sciences* (London: Kegan Paul, 1881).

16. N. Schwarz and F. Strack, 'Reports of Subjective Well-Being: Judgmental Processes and Their Methodological Implications', in *Well-Being: The Foundations of Hedonic Psychology*, ed. D. Kahneman, E. Diener and N. Schwarz (New York: Russell Sage Foundation, 1999), 61–84; D. Kahneman, 'Objective Happiness', in *Well-Being*, 3–25.

17. R. J. Larsen and B. L. Fredrickson, 'Measurement Issues in Emotion Research', in *Well-Being*, 40–60.

18. M. Minsky, *The Society of Mind* (New York: Simon & Schuster, 1985); W. G. Lycan, 'Homuncular Functionalism Meets PDP', in *Philosophy and Connectionist Theory*, ed. W. Ramsey, S. P. Stich and D. E. Rumelhart (Mahwah, NJ: Lawrence Erlbaum, 1991), 259–86.

19. O. Jowett, *Plato: Protagoras*, facsimile edn. (New York: Prentice Hall, 1956).

Chapter 4: In the Blind Spot of the Mind's Eye

1. R. O. Boyer and H. M. Morais, *Labor's Untold Story* (New York: Cameron, 1955); P. Avrich, *The Haymarket Tragedy* (Princeton, NJ: Princeton University Press, 1984).

2. E. Brayer, *George Eastman: A Biography* (Baltimore: Johns Hopkins University Press, 1996).

3. R. Karniol and M. Ross, 'The Motivational Impact of Temporal Focus: Thinking About the Future and the Past', *Annual Review of Psychology* 47: 593–620 (1996); and B. A. Mellers, 'Choice and the Relative Pleasure of Consequences', *Psychological Bulletin* 126: 910–24 (2000).

4. D. L. Schacter, *Searching for Memory: The Brain, the Mind and the Past* (New York: Basic Books, 1996).

5. E. F. Loftus, D. G. Miller and H. J. Burns, 'Semantic Integration of Verbal Information into Visual Memory', *Journal of Experimental Psychology: Human Learning and Memory* 4: 19–31 (1978).

6. E. F. Loftus, 'When a Lie Becomes Memory's Truth: Memory Distortion After Exposure to Misinformation', *Current Directions in Psychological Sciences* 1: 121–23 (1992). For an opposing view, see M. S. Zaragoza, M. McCloskey and M. Jamis, 'Misleading Postevent Information and Recall of the Original Event: Further Evidence Against the Memory Impairment Hypothesis', *Journal of Experimental Psychology: Learning, Memory, and Cognition* 13: 36–44 (1987).

7. M. K. Johnson and S. J. Sherman, 'Constructing and Reconstructing the Past and the Future in the Present', in *Handbook of Motivation and Cognition: Foundations of Social Behavior*, ed. E. T. Higgins and R. M. Sorrentino, vol. 2 (New York: Guilford Press, 1990), 482–526; and M. K. Johnson and C. L. Raye, 'Reality Monitoring', *Psychological Review* 88: 67–85 (1981).

8. J. Deese, 'On the Predicted Occurrence of Particular Verbal Intrusions in Immediate Recall', *Journal of Experimental Psychology* 58: 17–22 (1959).

9. H. L. Roediger and K. B. McDermott, 'Creating False Memories: Remembering Words Not Presented in Lists', *Journal of Experimental Psychology: Learning, Memory, and Cognition* 21: 803–14 (1995).

10. K. B. McDermott and H. L. Roediger, 'Attempting to Avoid Illusory Memories: Robust False Recognition of Associates Persists Under Conditions of Explicit Warnings and Immediate Testing,' *Journal of Memory and Language* 39: 508–20 (1998).

11. R. Warren, 'Perceptual Restoration of Missing Speech Sounds', *Science* 167: 392–93 (1970).

12. A. G. Samuel, 'A Further Examination of Attentional Effects in the Phonemic Restoration Illusion', *Quarterly Journal of Experimental Psychology* 43A: 679–99 (1991).

13. R. Warren, 'Perceptual Restoration of Obliterated Sounds', *Psychological Bulletin* 96: 371–83 (1984).

14. L. F. Baum, *The Wonderful Wizard of Oz* (New York: G. M. Hill, 1900), 113–19.

15. J. Locke, Book IV, *An Essay Concerning Human Understanding*, vol. 2 (1690; New York: Dover, 1959).

16. I. Kant, *Critique of Pure Reason*, trans. N. K. Smith (1781; New York: St. Martin's Press, 1965), 93.

17. W. Durant, *The Story of Philosophy* (New York: Simon & Schuster, 1926).

18. A. Gopnik and J. W. Astington, 'Children's Understanding of Representational Change and Its Relation to the Understanding of False Beliefs and the Appearance-Reality Distinction', *Child Development* 59: 26–37 (1988); and H. Wimmer and J. Perner, 'Beliefs About Beliefs: Representation and Constraining Function of Wrong Beliefs in Young Children's Understanding of Deception', *Cognition* 13: 103–28 (1983).

19. J. Piaget, *The Child's Conception of the World* (London: Routledge & Kegan Paul, 1929), 166.

20. B. Keysar et al., 'Taking Perspective in Conversation: The Role of Mutual Knowledge in Comprehension', *Psychological Science* 11: 32–38 (2000).

21. D. T. Gilbert, 'How Mental Systems Believe', *American Psychologist* 46: 107–19 (1991).

22. Interestingly, the ability to do this increases with age but begins to deteriorate in old age. See C. Ligneau-Hervé and E. Mullet, 'Perspective-Taking Judgments Among Young Adults, Middle-Aged, and Elderly People', *Journal of Experimental Psychology: Applied* 11: 53–60 (2005).

23. Piaget, *Child's Conception*, 124.

24. G. A. Miller, 'Trends and Debates in Cognitive Psychology', *Cognition* 10: 215–25 (1981).

25. D. T. Gilbert and T. D. Wilson, 'Miswanting: Some Problems in the Forecasting of Future Affective States', in *Feeling and Thinking: The Role of Affect in Social Cognition*, ed. J. Forgas (Cambridge: Cambridge University Press, 2000), 178–97.

26. D. Dunning et al., 'The Overconfidence Effect in Social Prediction', *Journal of Personality and Social Psychology* 58: 568–81 (1990); and R. Vallone et al., 'Overconfident Predictions of Future Actions and Outcomes by Self and Others', *Journal of Personality and Social Psychology* 58: 582–92 (1990).

27. D. W. Griffin, D. Dunning and L. Ross, 'The Role of Construal Processes in Overconfident Predictions About the Self and Others', *Journal of Personality and Social Psychology* 59: 1128–39 (1990).

28. Kant, *Critique*, 93.

Chapter 5: The Hound of Silence

1. A. C. Doyle, 'Silver Blaze', in *The Complete Sherlock Holmes* (1892; New York: Gramercy, 2002), 149.

2. Ibid.

3. R. S. Sainsbury and H. M. Jenkins, 'Feature-Positive Effect in Discrimination Learning', *Proceedings of the Annual Convention of the American Psychological Association* 2: 17–18 (1967).

4. J. P. Newman, W. T. Wolff and E. Hearst, 'The Feature-Positive Effect in Adult Human Subjects', *Journal of Experimental Psychology: Human Learning and Memory* 6: 630–50 (1980).

5. H. M. Jenkins and W. C. Ward, 'Judgment of Contingency Between Responses and Outcomes', *Psychological Monographs* 79 (1965); P. C. Wason, 'Reasoning About a Rule', *Quarterly Journal of Experimental Psychology* 20: 273–81 (1968); and D. L. Hamilton and R. K. Gifford, 'Illusory Correlation in Interpersonal Perception: A Cognitive Basis of Stereotypic Judgements', *Journal of Experimental Social Psychology* 12: 392–407 (1976). See also J. Crocker, 'Judgment of Covariation by Social Perceivers', *Psychological Bulletin* 90: 272–92 (1981); L. B. Alloy and N. Tabachnik, 'The Assessment of Covariation by Humans and Animals: The Joint Influence of Prior Expectations and Current Situational Information', *Psychological Review* 91: 112–49 (1984).

6. F. Bacon, *Novum organum*, ed. and trans. P. Urbach and J. Gibson (1620; Chicago: Open Court, 1994), 60.

7. Ibid., 57.

8. J. Klayman and Y. W. Ha, 'Confirmation, Disconfirmation, and Information in Hypothesis-Testing', *Psychological Review* 94: 211–28 (1987).

9. A. Tversky, 'Features of Similarity', *Psychological Review* 84: 327–52 (1977).

10. E. Shafir, 'Choosing Versus Rejecting: Why Some Options Are Both Better and Worse Than Others', *Memory & Cognition* 21: 546–56 (1993).

11. T. D. Wilson et al., 'Focalism: A Source of Durability Bias in Affective Forecasting', *Journal of Personality and Social Psychology* 78: 821–36 (2000).

12. This study was described in a paper that also described some other studies in which people were asked to make predictions about how they would feel if *(a)* the space shuttle *Columbia* exploded and killed all the astronauts on board, or *(b)* an American-led war in Iraq deposed Saddam Hussein. The spooky thing is

that the studies were conducted in 1998 – five years before either of these events actually took place. Believe it or not.

13. D. A. Schkade and D. Kahneman, 'Does Living in California Make People Happy? A Focusing Illusion in Judgments of Life Satisfaction', *Psychological Science* 9: 340–46 (1998).

14. This may be less true of people who live in cultures that emphasize holistic thinking. See K. C. H. Lam et al., 'Cultural Differences in Affective Forecasting: The Role of Focalism', *Personality and Social Psychology Bulletin* 31: 1296–309 (2005).

15. P. Menzela et al., 'The Role of Adaptation to Disability and Disease in Health State Valuation: A Preliminary Normative Analysis', *Social Science & Medicine* 55: 2149–58 (2002).

16. C. Turnbull, *The Forest People* (New York: Simon & Schuster, 1961), 222.

17. Y. Trope and N. Liberman, 'Temporal Construal', *Psychological Review* 110: 403–21 (2003).

18. R. R. Vallacher and D. M. Wegner, *A Theory of Action Identification* (Hillsdale, NJ: Lawrence Erlbaum, 1985), 61–88.

19. N. Liberman and Y. Trope, 'The Role of Feasibility and Desirability Considerations in Near and Distant Future Decisions: A Test of Temporal Construal Theory', *Journal of Personality and Social Psychology* 75: 5–18 (1998).

20. M. D. Robinson and G. L. Clore, 'Episodic and Semantic Knowledge in Emotional Self-Report: Evidence for Two Judgment Processes', *Journal of Personality and Social Psychology* 83: 198–215 (2002).

21. T. Eyal et al., 'The Pros and Cons of Temporally Near and Distant Action', *Journal of Personality and Social Psychology* 86: 781–95 (2004).

22. I. R. Newby-Clark and M. Ross, 'Conceiving the Past and Future', *Personality and Social Psychology Bulletin* 29: 807–18 (2003); and M. Ross and I. R. Newby-Clark, 'Construing the Past and Future', *Social Cognition* 16: 133–50 (1998).

23. N. Liberman, M. Sagristano and Y. Trope, 'The Effect of Temporal Distance on Level of Mental Construal', *Journal of Experimental Social Psychology* 38: 523–34 (2002).

24. G. Ainslie, 'Specious Reward: A Behavioral Theory of Impulsiveness and Impulse Control', *Psychological Bulletin* 82: 463–96 (1975); and G. Ainslie, *Picoeconomics: The Strategic Interaction of Successive Motivational States Within the Person* (Cambridge: Cambridge University Press, 1992).

25. The first author to notice this was Plato, who used this fact to make his case for an objective measurement of happiness: 'Do not the same magnitudes appear larger to your sight when near, and smaller when at a distance? . . . Now suppose happiness to consist in doing or choosing the greater, and in not doing or in avoiding the less, what would be the saving principle of human life? Would not the art of measuring be the saving principle; or would the power of appearance? Is not the latter that deceiving art which makes us wander up and down and take the things at one time of which we repent at another, both in our actions and in our choice of things great and small?' O. Jowett, *Plato: Protagoras*, facsimile edn. (New York: Prentice Hall, 1956).

26. G. Loewenstein, 'Anticipation and the Valuation of Delayed Consumption', *Economy Journal* 97: 666–84 (1987).

27. S. M. McClure et al., 'The Grasshopper and the Ant: Separate Neural Systems Value Immediate and Delayed Monetary Rewards', *Science* 306: 503–7 (2004).

28. Doyle, *Complete Sherlock Holmes*, 147.

Chapter 6: The Future Is Now

1. *Everyone* seems sure that Kelvin said this in 1895, but for the life of me I can't find the original source to prove it.

2. S. A. Newcomb, *Side-Lights on Astronomy* (New York: Harper & Brothers, 1906), 355.

3. W. Wright, 'Speech to the Aero Club of France', in *The Papers of Wilbur and Orville Wright*, ed. M. McFarland (New York: McGraw-Hill, 1908), 934.

4. A. C. Clarke, *Profiles of the Future* (New York: Bantam, 1963), 14. By the way, Clarke defines 'elderly' as somewhere between thirty and forty-five. Yikes!

5. G. R. Goethals and R. F. Reckman, 'The Perception of Consistency in Attitudes', *Journal of Experimental Social Psychology* 9: 491–501 (1973).

6. C. McFarland and M. Ross, 'The Relation Between Current Impressions and Memories of Self and Dating Partners', *Personality and Social Psychology Bulletin* 13: 228–38 (1987).

7. M. A. Safer, L. J. Levine and A. L. Drapalski, 'Distortion in Memory for Emotions: The Contributions of Personality and Post-Event Knowledge', *Personality and Social Psychology Bulletin* 28: 1495–1507 (2002).

8. E. Eich et al., 'Memory for Pain: Relation Between Past and Present Pain Intensity', *Pain* 23: 375–80 (1985).

9. L. N. Collins et al., 'Agreement Between Retrospective Accounts of Substance Use and Earlier Reported Substance Use', *Applied Psychological Measurement* 9: 301–9 (1985); G. B. Markus, 'Stability and Change in Political Attitudes: Observe, Recall, and "Explain" *Political Behavior* 8: 21–44 (1986); D. Offer et al., 'The Altering of Reported Experiences', *Journal of the American Academy of Child and Adolescent Psychiatry* 39: 735–42 (2000).

10. M. A. Safer, G. A. Bonanno and N. P. Field, '"It Was Never That Bad": Biased Recall of Grief and Long-Term Adjustment to the Death of a Spouse', *Memory* 9: 195–204 (2001).

11. For reviews, see M. Ross, 'Relation of Implicit Theories to the Construction of Personal Histories', *Psychological Review* 96: 341–57 (1989); L. J. Levine and M. A. Safer, 'Sources of Bias in Memory for Emotions', *Current Directions in Psychological Science* 11: 169–73 (2002).

12. L. J. Levine, 'Reconstructing Memory for Emotions', *Journal of Experimental Psychology: General* 126: 165–77 (1997).

13. G. F. Loewenstein, 'Out of Control: Visceral Influences on Behavior', *Organizational Behavior and Human Decision Processes* 65: 272–92 (1996); G. F. Loewenstein, T. O'Donoghue and M. Rabin, 'Projection Bias in Predicting Future Utility', *Quarterly Journal of Economics* 118: 1209–48 (2003);

G. Loewenstein and E. Angner, 'Predicting and Indulging Changing Preferences', in *Time and Decision*, ed. G. Loewenstein, D. Read and R. F. Baumeister (New York: Russell Sage Foundation, 2003), 351–91; L. van Boven, D. Dunning and G. F. Loewenstein, 'Egocentric Empathy Gaps Between Owners and Buyers: Misperceptions of the Endowment Effect', *Journal of Personality and Social Psychology* 79: 66–76 (2000).

14. R. E. Nisbett and D. E. Kanouse, 'Obesity, Food Deprivation and Supermarket Shopping Behavior', *Journal of Personality and Social Psychology* 12: 289–94 (1969); D. Read and B. van Leeuwen, 'Predicting Hunger: The Effects of Appetite and Delay on Choice', *Organizational Behavior and Human Decision Processes* 76: 189–205 (1998).

15. G. F. Loewenstein, D. Prelec and C. Shatto, 'Hot/Cold Intrapersonal Empathy Gaps and the Under-prediction of Curiosity' (unpublished manuscript, Carnegie-Mellon University 1998), cited in G. F. Loewenstein, 'The Psychology of Curiosity: A Review and Reinterpretation', *Psychological Bulletin* 116: 75–98 (1994).

16. S. M. Kosslyn et al., 'The Role of Area 17 in Visual Imagery: Convergent Evidence from PET and rTMS', *Science* 284: 167–70 (1999).

17. P. K. McGuire, G. M. S. Shah and R. M. Murray, 'Increased Blood Flow in Broca's Area During Auditory Hallucinations in Schizophrenia', *Lancet* 342: 703–6 (1993).

18. D. J. Kavanagh, J. Andrade and J. May, 'Imaginary Relish and Exquisite Torture: The Elaborated Intrusion Theory of Desire', *Psychological Review* 112: 446–67 (2005).

19. A. K. Anderson and E. A. Phelps, 'Lesions of the Human Amygdala Impair Enhanced Perception of Emotionally Salient Events', *Nature* 411: 305–9 (2001); E. A. Phelps et al., 'Activation of the Left Amygdala to a Cognitive Representation of Fear', *Nature Neuroscience* 4: 437–41 (2001); and H. C. Breiter et al., 'Functional Imaging of Neural Responses to Expectancy and Experience of Monetary Gains and Losses', *Neuron* 30 (2001).

20. As far as I can tell, the word *prefeel* was first used as a song title on the 1999 album *Prize*, by Arto Lindsay. See also C. M. Atance and D. K. O'Neill, 'Episodic Future Thinking', *Trends in Cognitive Sciences* 5: 533–39 (2001).

21. T. D. Wilson et al., 'Introspecting About Reasons Can Reduce Post-Choice Satisfaction', *Personality and Social Psychology Bulletin* 19: 331–39 (1993). See also T. D. Wilson and J. W. Schooler, 'Thinking Too Much: Introspection Can Reduce the Quality of Preferences and Decisions', *Journal of Personality and Social Psychology* 60: 181–92 (1991).

22. C. N. DeWall and R. F. Baumeister, 'Alone but Feeling No Pain: Effects of Social Exclusion on Physical Pain Tolerance and Pain Threshold, Affective Forecasting, and Interpersonal Empathy', *Journal of Personality and Social Psychology* (in press).

23. D. Reisberg et al., '"Enacted" Auditory Images Are Ambiguous; "Pure" Auditory Images Are Not', *Quarterly Journal of Experimental Psychology: Human Experimental Psychology* 41: 619–41 (1989).

24. Clever psychologists have been able to design some unusual circum-

stances that provide exceptions to this rule; see C. W. Perky, 'An Experimental Study of Imagination', *American Journal of Psychology* 21: 422–52 (1910). It is also worth noting that while we can almost always distinguish between what we are seeing and what we are imagining, we are not always able to distinguish between what we saw and what we imagined; see M. K. Johnson and C. L. Raye, 'Reality Monitoring', *Psychological Review* 88: 67–85 (1981).

25. N. Schwarz and G. L. Clore, 'Mood, Misattribution, and Judgments of Well-Being: Informative and Directive Functions of Affective States', *Journal of Personality and Social Psychology* 45: 513–23 (1983).

26. L. van Boven and G. Loewenstein, 'Social Projection of Transient Drive States', *Personality and Social Psychology Bulletin* 29: 1159–68 (2003).

27. A. K. MacLeod and M. L. Cropley, 'Anxiety, Depression, and the Anticipation of Future Positive and Negative Experiences', *Journal of Abnormal Psychology* 105: 286–89 (1996).

28. E. J. Johnson and A. Tversky, 'Affect, Generalization, and the Perception of Risk', *Journal of Personality and Social Psychology* 45: 20–31 (1983); and D. DeSteno et al., 'Beyond Valence in the Perception of Likelihood: The Role of Emotion Specificity', *Journal of Personality and Social Psychology* 78: 397–416 (2000).

Chapter 7: Time Bombs

1. M. Hegarty, 'Mechanical Reasoning by Mental Simulation', *Trends in Cognitive Sciences* 8: 280–85 (2004).

2. G. Lakoff and M. Johnson, *Metaphors We Live By* (Chicago: University of Chicago Press, 1980).

3. D. Gentner, M. Imai and L. Boroditsky, 'As Time Goes By: Evidence for Two Systems in Processing Space Time Metaphors', *Language and Cognitive Processes* 17: 537–65 (2002); and L. Boroditsky, 'Metaphoric Structuring: Understanding Time Through Spatial Metaphors', *Cognition* 75: 1–28 (2000).

4. B. Tversky, S. Kugelmass and A. Winter, 'Cross-Cultural and Developmental Trends in Graphic Productions', *Cognitive Psychology* 23: 515–57 (1991).

5. L. Boroditsky, 'Does Language Shape Thought? Mandarin and English Speakers' Conceptions of Time', *Cognitive Psychology* 43: 1–22 (2001).

6. R. K. Ratner, B. E. Kahn and D. Kahneman, 'Choosing Less-Preferred Experiences for the Sake of Variety', *Journal of Consumer Research* 26: 1–15 (1999).

7. D. Read and G. F. Loewenstein, 'Diversification Bias: Explaining the Discrepancy in Variety Seeking Between Combined and Separated Choices', *Journal of Experimental Psychology: Applied* 1: 34–49 (1995). See also I. Simonson, 'The Effect of Purchase Quantity and Timing on Variety-Seeking Behavior', *Journal of Marketing Research* 27: 150–62 (1990).

8. T. D. Wilson and D. T. Gilbert, 'Making Sense: A Model of Affective Adaptation' (unpublished manuscript, University of Virginia, 2005).

9. Human beings are not the only animals that appreciate variety. The

Coolidge effect ostensibly got its name when President Calvin Coolidge and his wife were touring a farm. The foreman noted the sexual prowess of his prize rooster: 'This rooster can have sex all day without stopping,' he said. 'Really?' said Mrs. Coolidge. 'Please tell that to my husband.' The president turned to the foreman and asked, 'Does the rooster mate with the same chicken each time?' 'No,' said the foreman, 'always with a different chicken.' To which the president replied, 'Really? Please tell *that* to my wife.' The story is probably apocryphal, but the phenomenon is not: male mammals who have mated to exhaustion can usually be induced to mate again with a novel female; see J. Wilson, R. Kuehn and F. Beach, 'Modifications in the Sexual Behavior of Male Rats Produced by Changing the Stimulus Female', *Journal of Comparative and Physiological Psychology* 56: 636–44 (1963). In fact, even breeding bulls whose sperm is collected by a machine show a greatly reduced time to ejaculation when the machine to which they've become habituated is moved to a novel location. E. B. Hale and J. O. Almquist, 'Relation of Sexual Behavior to Germ Cell Output in Farm Animals', *Journal of Dairy Science* 43: Supp., 145–67 (1960).

10. It is worth noting that if we change our assumptions (particularly the habituation assumption), then switching plates exactly at the point at which our favourite meal has lost its edge can be a poor strategy for maximizing pleasure in the long term; R. J. Hernstein, *The Matching Law: Papers in Psychology and Economics*, ed. H. Rachlin and D. I. Laibson (Cambridge, Mass.: Harvard University Press, 1997).

11. I would like to take this opportunity to note that while my apparent fixation on adulterous mailmen betrays my juvenile sense of humour, the example is entirely fictitious and is not meant to reflect poorly on the many fine mailmen and spouses I have had.

12. D. T. Gilbert, 'Inferential Correction', in *Heuristics and Biases: The Psychology of Intuitive Judgment*, ed. T. Gilovich, D. W. Griffin and D. Kahneman (Cambridge: Cambridge University Press, 2002), 167–84.

13. A. Tversky and D. Kahneman, 'Judgment Under Uncertainty: Heuristics and Biases', *Science* 185: 1124–31 (1974).

14. N. Epley and T. Gilovich, 'Putting Adjustment Back in the Anchoring and Adjustment Heuristic: Differential Processing of Self-Generated and Experimenter-Provided Anchors', *Psychological Science* 12: 391–96 (2001).

15. D. T. Gilbert, M. J. Gill and T. D. Wilson, 'The Future Is Now: Temporal Correction in Affective Forecasting,' *Organizational Behavior and Human Decision Processes* 88: 430–44 (2002).

16. See also J. E. J. Ebert, 'The Role of Cognitive Resources in the Valuation of Near and Far Future Events', *Acta Psychologica* 108: 155–71 (2001).

17. G. F. Loewenstein and D. Prelec, 'Preferences for Sequences of Outcomes', *Psychological Review* 100: 91–108 (1993).

18. D. Kahneman and A. Tversky, 'Prospect Theory: An Analysis of Decision Under Risk', *Econometrica* 47: 263–91 (1979).

19. J. W. Pratt, D. A. Wise and R. Zeckhauser, 'Price Differences in Almost Competitive Markets', *Quarterly Journal of Economics* 93: 189–211 (1979); A. Tversky and D. Kahneman, 'The Framing of Decisions and the Psychology of

Choice', *Science* 211: 453–58 (1981); R. H. Thaler, 'Toward a Positive Theory of Consumer Choice', *Journal of Economic Behavior and Organization* 1: 39–60 (1980).

20. R. H. Thaler, 'Mental Accounting Matters', *Journal of Behavioral Decision Making* 12: 183–206 (1999).

21. R. B. Cialdini et al., 'Reciprocal Concessions Procedure for Inducing Compliance: The Door-in-the-Face Technique', *Journal of Personality and Social Psychology* 31: 206–15 (1975). There is some controversy about whether this effect is, in fact, due to the contrast between the large and small requests. See J. P. Dillard, 'The Current Status of Research on Sequential-Request Compliance Techniques', *Personality and Social Psychology Bulletin* 17: 283–88 (1991).

22. D. Kahneman and D. T. Miller, 'Norm Theory: Comparing Reality to Its Alternatives', *Psychological Review* 93: 136–53 (1986).

23. O. E. Tykocinski and T. S. Pittman, 'The Consequences of Doing Nothing: Inaction Inertia as Avoidance of Anticipated Counterfactual Regret', *Journal of Personality and Social Psychology* 75: 607–16 (1998); and O. E. Tykocinski, T. S. Pittman, and E. E. Tuttle, 'Inaction Inertia: Forgoing Future Benefits as a Result of an Initial Failure to Act', *Journal of Personality and Social Psychology* 68: 793–803 (1995).

24. D. Kahneman and A. Tversky, 'Choices, Values, and Frames', *American Psychologist* 39: 341–50 (1984).

25. I. Simonson and A. Tversky, 'Choice in Context: Tradeoff Contrast and Extremeness Aversion', *Journal of Marketing Research* 29: 281–95 (1992).

26. R. B. Cialdini, *Influence: Science and Practice* (Glenview, Ill.: Scott, Foresman, 1985).

27. D. A. Redelmeier and E. Shafir, 'Medical Decision Making in Situations That Offer Multiple Alternatives', *JAMA: Journal of the American Medical Association* 273: 302–5 (1995).

28. S. S. Iyengar and M. R. Lepper, 'When Choice Is Demotivating: Can One Desire Too Much of a Good Thing?' *Journal of Personality and Social Psychology* 79: 995–1006 (2000); and B. Schwartz, 'Self-Determination: The Tyranny of Freedom', *American Psychologist* 55: 79–88 (2000).

29. A. Tversky, S. Sattath and P. Slovic, 'Contingent Weighting in Judgment and Choice,' *Psychological Review* 95: 371–84 (1988).

30. C. K. Hsee et al., 'Preference Reversals Between Joint and Separate Evaluations of Options: A Review and Theoretical Analysis', *Psychological Bulletin* 125: 576–90 (1999).

31. C. Hsee, 'The Evaluability Hypothesis: An Explanation for Preference Reversals Between Joint and Separate Evaluations of Alternatives', *Organizational Behavior and Human Decision Processes* 67: 247–57 (1996).

32. J. R. Priester, U. M. Dholakia and M. A. Fleming, 'When and Why the Background Contrast Effect Emerges: Thought Engenders Meaning by Influencing the Perception of Applicability', *Journal of Consumer Research* 31: 491–501 (2004).

33. K. Myrseth, C. K. Morewedge and D. T. Gilbert, unpublished raw data, Harvard University, 2004.

34. D. Kahneman and A. Tversky, 'Prospect Theory: An Analysis of Decision Under Risk', *Econometrica* 47: 263–91 (1979); A. Tversky and D. Kahneman, 'The Framing of Decisions and the Psychology of Choice', *Science* 211: 453–58 (1981); and A. Tversky and D. Kahneman, 'Loss Aversion in Riskless Choice: A Reference-Dependent Model', *Quarterly Journal of Economics* 106: 1039–61 (1991).

35. D. Kahneman, J. L. Knetsch and R. H. Thaler, 'Experimental Tests of the Endowment Effect and the Coase Theorem', *Journal of Political Economy* 98: 1325–48 (1990); and D. Kahneman, J. Kentsch and D. Thaler, 'The Endowment Effect, Loss Aversion, and Status Quo Bias', *Journal of Economic Perspectives* 5: 193–206 (1991).

36. L. van Boven, D. Dunning and G. F. Loewenstein, 'Egocentric Empathy Gaps Between Owners and Buyers: Misperceptions of the Endowment Effect', *Journal of Personality and Social Psychology* 79: 66–76 (2000); and Z. Carmon and D. Ariely, 'Focusing on the Foregone: How Value Can Appear So Different to Buyers and Sellers', *Journal of Consumer Research* 27: 360–70 (2000).

37. L. Hunt, 'Against Presentism', *Perspectives* 40 (2002).

Chapter 8: Paradise Glossed

1. C. B. Wortman and R. C. Silver, 'The Myths of Coping with Loss', *Journal of Consulting and Clinical Psychology* 57: 349–57 (1989); G. A. Bonanno, 'Loss, Trauma, and Human Resilience: Have We Underestimated the Human Capacity to Thrive After Extremely Aversive Events?', *American Psychologist* 59: 20–28 (2004); and C. S. Carver, 'Resilience and Thriving: Issues, Models, and Linkages', *Journal of Social Issues* 54: 245–66 (1998).

2. G. A. Bonanno and S. Kaltman, 'Toward an Integrative Perspective on Bereavement', *Psychological Bulletin* 125: 760–76 (1999); and G. A. Bonanno et al., 'Resilience to Loss and Chronic Grief: A Prospective Study from Preloss to 18-Months Postloss', *Journal of Personality and Social Psychology* 83: 1150–64 (2002).

3. E. J. Ozer et al., 'Predictors of Posttraumatic Stress Disorder and Symptoms in Adults: A Meta-analysis', *Psychological Bulletin* 129: 52–73 (2003).

4. G. A. Bonanno, C. Rennicke and S. Dekel, 'Self-Enhancement Among High-Exposure Survivors of the September 11th Terrorist Attack: Resilience or Social Maladjustment?', *Journal of Personality and Social Psychology* 88: 984–98 (2005).

5. R. G. Tedeschi and L. G. Calhoun, 'Posttraumatic Growth: Conceptual Foundations and Empirical Evidence', *Psychological Inquiry* 15: 1–18 (2004); P. A. Linley and S. Joseph, 'Positive Change Following Trauma and Adversity: A Review', *Journal of Traumatic Stress* 17: 11–21 (2004); and C. S. Carver, 'Resilience and Thriving: Issues, Models, and Linkages', *Journal of Social Issues* 54: 245–66 (1998).

6. K. Sack, 'After 37 Years in Prison, Inmate Tastes Freedom', *New York Times*, 11 January 1996, 18.

7. C. Reeve, Ohio State University commencement speech, 13 June 2003.

8. D. Becker, 'Cycling Through Adversity: Ex-World Champ Stays on Cancer Comeback Course', *USA Today*, 22 May 1998, 3C.

9. R. G. Tedeschi and L. G. Calhoun, *Trauma and Transformation: Growing in the Aftermath of Suffering* (Sherman Oaks, Calif.: Sage, 1995), 1.

10. R. Schulz and S. Decker, 'Long-Term Adjustment to Physical Disability: The Role of Social Support, Perceived Control, and Self-Blame', *Journal of Personality and Social Psychology* 48: 1162–72 (1985); C. B. Wortman and R. C. Silver, 'Coping with Irrevocable Loss', in *Cataclysms, Crises, and Catastrophes: Psychology in Action*, ed. G. R. VandenBos and B. K. Bryant (Washington, DC: American Psychological Association, 1987), 185–235; and P. Brickman, D. Coates and R. J. Janoff-Bulman, 'Lottery Winners and Accident Victims: Is Happiness Relative?', *Journal of Personality and Social Psychology* 36: 917–27 (1978).

11. S. E. Taylor, 'Adjustment to Threatening Events: A Theory of Cognitive Adaptation', *American Psychologist* 38: 1161–73 (1983).

12. D. T. Gilbert, E. Driver-Linn and T. D. Wilson, 'The Trouble with Vronsky: Impact Bias in the Forecasting of Future Affective States', in *The Wisdom in Feeling: Psychological Processes in Emotional Intelligence*, ed. L. F. Barrett and P. Salovey (New York: Guilford Press, 2002), 114–43; and T. D. Wilson and D. T. Gilbert, 'Affective Forecasting', in *Advances in Experimental Social Psychology*, ed. M. Zanna, vol. 35 (New York: Elsevier, 2003).

13. D. L. Sackett and G. W. Torrance, 'The Utility of Different Health States as Perceived by the General Public', *Journal of Chronic Disease* 31: 697–704 (1978); P. Dolan and D. Kahneman, 'Interpretations of Utility and Their Implications for the Valuation of Health' (unpublished manuscript, Princeton University, 2005); and J. Riis et al., 'Ignorance of Hedonic Adaptation to Hemo-Dialysis: A Study Using Ecological Momentary Assessment', *Journal of Experimental Psychology: General* 134: 3–9 (2005).

14. P. Menzela et al., 'The Role of Adaptation to Disability and Disease in Health State Valuation: A Preliminary Normative Analysis', *Social Science & Medicine* 55: 2149–58 (2002).

15. P. Dolan, 'Modelling Valuations for EuroQol Health States', *Medical Care* 11: 1095–1108 (1997).

16. J. Jonides and H. Gleitman, 'A Conceptual Category Effect in Visual Search: O as Letter or as Digit', *Perception and Psychophysics* 12: 457–60 (1972).

17. C. M. Solley and J. F. Santos, 'Perceptual Learning with Partial Verbal Reinforcement', *Perceptual and Motor Skills 8*: 183–93 (1958); and E. D. Turner and W. Bevan, 'Patterns of Experience and the Perceived Orientation of the Necker Cube', *Journal of General Psychology* 70: 345–52 (1964).

18. D. Dunning, J. A. Meyerowitz and A. D. Holzberg, 'Ambiguity and Self-Evaluation: The Role of Idiosyncratic Trait Definitions in Self-Serving Assessments of Ability', *Journal of Personality and Social Psychology* 57: 1–9 (1989).

19. C. K. Morewedge and D. T. Gilbert, unpublished raw data, Harvard University, 2004.

20. J. W. Brehm, 'Post-decision Changes in Desirability of Alternatives', *Journal of Abnormal and Social Psychology* 52: 384–89 (1956).

21. E. E. Lawler et al., 'Job Choice and Post Decision Dissonance', *Organizational Behavior and Human Decision Processes* 13: 133–45 (1975).

22. S. Lyubomirsky and L. Ross, 'Changes in Attractiveness of Elected, Rejected, and Precluded Alternatives: A Comparison of Happy and Unhappy Individuals', *Journal of Personality and Social Psychology* 76: 988–1007 (1999).

23. R. E. Knox and J. A. Inkster, 'Postdecision Dissonance at Post Time', *Journal of Personality and Social Psychology* 8: 319–23 (1968).

24. O. J. Frenkel and A. N. Doob, 'Post-decision Dissonance at the Polling Booth', *Canadian Journal of Behavioural Science* 8: 347–50 (1976).

25. F. M. Voltaire, *Candide* (1759), chap. 1. I've never found an English translation that I like, so I've cobbled together my own. Pardon my French.

26. R. F. Baumeister, 'The Optimal Margin of Illusion', *Journal of Social and Clinical Psychology* 8: 176–89 (1989); S. E. Taylor, *Positive Illusions* (New York: Basic Books, 1989); S. E. Taylor and J. D. Brown, 'Illusion and Well-Being: A Social-Psychological Perspective on Mental Health', *Psychological Bulletin* 103: 193–210 (1988); Z. Kunda, 'The Case for Motivated Reasoning', *Psychological Bulletin* 108: 480–98 (1990); and T. Pyszczynski and J. Greenberg, 'Toward an Integration of Cognitive and Motivational Perspectives on Social Inference: A Biased Hypothesis-Testing Model', in *Advances in Experimental Social Psychology*, ed. L. E. Berkowitz, vol. 20 (San Diego: Academic Press, 1987), 297–340.

27. Both Sigmund and Anna Freud called this a system of 'defence mechanisms', and just about every psychologist since has remarked on that system and given it a different name. A recent summary of the literature on psychological defence can be found in D. L. Paulhus, B. Fridhandler and S. Hayes, 'Psychological Defense: Contemporary Theory and Research', in *Handbook of Personality Psychology*, ed. R. Hogan, J. Johnson and S. Briggs (San Diego: Academic Press, 1997), 543–79.

28. W. B. Swann, B. W. Pelham and D. S. Krull, 'Agreeable Fancy or Disagreeable Truth? Reconciling Self-Enhancement and Self-Verification', *Journal of Personality and Social Psychology* 57: 782–91 (1989); W. B. Swann, P. J. Rentfrow and J. Guinn, 'Self-Verification: The Search for Coherence', in *Handbook of Self and Identity*, ed. M. Leary and J. Tagney (New York: Guilford Press, 2002), 367–83; and W. B. Swann, *Self-Traps: The Elusive Quest for Higher Self-Esteem* (New York: Freeman, 1996).

29. W. B. Swann and B. W. Pelham, 'Who Wants Out When the Going Gets Good? Psychological Investment and Preference for Self-Verifying College Roommates', *Journal of Self and Identity* 1: 219–33 (2002).

30. J. L. Freedman and D. O. Sears, 'Selective Exposure', in *Advances in Experimental Social Psychology*, ed. L. Berkowitz, vol. 2 (New York: Academic Press, 1965), 57–97; and D. Frey, 'Recent Research on Selective Exposure to Information', in *Advances in Experimental Social Psychology*, ed. L. Berkowitz, vol. 19 (New York: Academic Press, 1986), 41–80.

31. D. Frey and D. Stahlberg, 'Selection of Information After Receiving More or Less Reliable Self-Threatening Information', *Personality and Social Psychology Bulletin* 12: 434–41 (1986).

32. B. Holton and T. Pyszczynski, 'Biased Information Search in the Interpersonal Domain', *Personality and Social Psychology Bulletin* 15: 42–51 (1989).

33. D. Ehrlich et al., 'Postdecision Exposure to Relevant Information', *Journal of Abnormal and Social Psychology* 54: 98–102 (1957).

34. R. Sanitioso, Z. Kunda and G. T. Fong, 'Motivated Recruitment of Autobiographical Memories', *Journal of Personality and Social Psychology* 59: 229–41 (1990).

35. A. Tesser and S. Rosen, 'Similarity of Objective Fate as a Determinant of the Reluctance to Transmit Unpleasant Information: The MUM Effect', *Journal of Personality and Social Psychology* 23: 46–53 (1972).

36. M. Snyder and W. B. Swann, Jr., 'Hypothesis Testing Processes in Social Interaction', *Journal of Personality and Social Psychology* 36: 1202–12 (1978); and W. B. J. Swann, T. Giuliano and D. M. Wegner, 'Where Leading Questions Can Lead: The Power of Conjecture in Social Interaction', *Journal of Personality and Social Psychology* 42: 1025–35 (1982).

37. D. T. Gilbert and E. E. Jones, 'Perceiver-Induced Constraint: Interpretations of Self-Generated Reality', *Journal of Personality and Social Psychology* 50: 269–80 (1986).

38. L. Festinger, 'A Theory of Social Comparison Processes', *Human Relations* 7: 117–40 (1954); A. Tesser, M. Millar and J. Moore, 'Some Affective Consequences of Social Comparison and Reflection Processes: The Pain and Pleasure of Being Close', *Journal of Personality and Social Psychology* 54: 49–61 (1988); S. E. Taylor and M. Lobel, 'Social Comparison Activity Under Threat: Downward Evaluation and Upward Contacts', *Psychological Review* 96: 569–75 (1989); and T. A. Wills, 'Downward Comparison Principles in Social Psychology', *Psychological Bulletin* 90: 245–71 (1981).

39. T. Pyszczynski, J. Greenberg and J. LaPrelle, 'Social Comparison After Success and Failure: Biased Search for Information Consistent with a Self-Servicing Conclusion', *Journal of Experimental Social Psychology* 21: 195–211 (1985).

40. J. V. Wood, S. E. Taylor and R. R. Lichtman, 'Social Comparison in Adjustment to Breast Cancer', *Journal of Personality and Social Psychology* 49: 1169–83 (1985).

41. S. E. Taylor et al., 'Social Support, Support Groups, and the Cancer Patient', *Journal of Consulting and Clinical Psychology* 54: 608–15 (1986).

42. A. Tesser and J. Smith, 'Some Effects of Task Relevance and Friendship on Helping: You Don't Always Help the One You Like' *Journal of Experimental Social Psychology* 16: 582–90 (1980).

43. A. H. Hastorf and H. Cantril, 'They Saw a Game: A Case Study' *Journal of Abnormal and Social Psychology* 49: 129–34 (1954).

44. L. Sigelman and C. K. Sigelman, 'Judgments of the Carter-Reagan Debate: The Eyes of the Beholders', *Public Opinion Quarterly* 48: 624–28 (1984); R. K. Bothwell and J. C. Brigham, 'Selective Evaluation and Recall During the 1980 Reagan-Carter Debate', *Journal of Applied Social Psychology* 13: 427–42 (1983); J. G. Payne et al., 'Perceptions of the 1988 Presidential and Vice-Presidential Debates', *American Behavioral Scientist* 32: 425–35 (1989);

and G. D. Munro et al., 'Biased Assimilation of Sociopolitical Arguments: Evaluating the 1996 U.S. Presidential Debate', *Basic and Applied Social Psychology* 24: 15–26 (2002).

45. R. P. Vallone, L. Ross and M. R. Lepper, 'The Hostile Media Phenomenon: Biased Perception and Perceptions of Media Bias in Coverage of the Beirut Massacre', *Journal of Personality and Social Psychology* 49: 577–85 (1985).

46. C. G. Lord, L. Ross and M. R. Lepper, 'Biased Assimilation and Attitude Polarization: The Effects of Prior Theories on Subsequently Considered Evidence', *Journal of Personality and Social Psychology* 37: 2098–109 (1979).

47. It is no consolation that in subsequent studies, both established scientists and scientists in training showed the same tendency to favour techniques that produced favoured conclusions. See J. J. Koehler, 'The Influence of Prior Beliefs on Scientific Judgments of Evidence Quality', *Organizational Behavior and Human Decision Processes* 56: 28–55 (1993).

48. T. Pyszczynski, J. Greenberg and K. Holt, 'Maintaining Consistency Between Self-Serving Beliefs and Available Data: A Bias in Information Evaluation', *Personality and Social Psychology Bulletin* 11: 179–90 (1985).

49. P. H. Ditto and D. F. Lopez, 'Motivated Skepticism: Use of Differential Decision Criteria for Preferred and Nonpreferred Conclusions', *Journal of Personality and Social Psychology* 63: 568–84 (1992).

50. Ibid.

51. T. Gilovich, *How We Know What Isn't So: The Fallibility of Human Reason in Everyday Life* (New York: Free Press, 1991).

52. This tendency can have disastrous consequences. For example, in 2004, the U.S. Senate Intelligence Committee concluded that the CIA had provided the White House with incorrect information about Iraq's weapons of mass destruction, which led the United States to invade Iraq. According to that report, the tendency to cook facts 'led Intelligence Community analysts, collectors and managers to both interpret ambiguous evidence as conclusively indicative of a WMD program as well as ignore or minimize evidence that Iraq did not have active and expanding weapons of mass destruction programs'. K. P. Shrader, 'Report: War Rationale Based on CIA Error', Associated Press, 9 July 2004.

53. Agence-France-Presse, 'Italy: City Wants Happier Goldfish', *New York Times*, 24 July 2004, A5.

Chapter 9: Immune to Reality

1. T. D. Wilson, *Strangers to Ourselves: Discovering the Adaptive Unconscious* (Cambridge, Mass.: Harvard University Press, 2002); and J. A. Bargh and T. L. Chartrand, 'The Unbearable Automaticity of Being', *American Psychologist* 54: 462–79 (1999).

2. R. E. Nisbett and T. D. Wilson, 'Telling More Than We Can Know: Verbal Reports on Mental Processes', *Psychological Review* 84: 231–59 (1977); D. J. Bem, 'Self-Perception Theory', in *Advances in Experimental Social Psychology*, ed. L. Berkowitz, vol. 6 (New York: Academic Press, 1972), 1–62; M. S. Gazzaniga, *The Social Brain* (New York: Basic Books, 1985); and

D. M. Wegner, *The Illusion of Conscious Will* (Cambridge, Mass.: MIT Press, 2003).

3. E. T. Higgings, W. S. Rholes and C. R. Jones, 'Category Accessibility and Impression Formation', *Journal of Experimental Social Psychology* 13: 141–54 (1977).

4. J. Bargh, M. Chen and L. Burrows 'Automaticity of Social Behavior: Direct Effects of Trait Construct and Stereotype Activation on Action', *Journal of Personality and Social Psychology* 71: 230–44 (1996).

5. A. Dijksterhuis and A. van Knippenberg, 'The Relation Between Perception and Behavior, or How to Win a Game of Trivial Pursuit', *Journal of Personality and Social Psychology* 74: 865–77 (1998).

6. Nisbett and Wilson, 'Telling More Than We Can Know'.

7. J. W. Schooler, D. Ariely and G. Loewenstein, 'The Pursuit and Assessment of Happiness Can Be Self-Defeating', in *The Psychology of Economic Decisions: Rationality and Well-Being*, ed. I. Brocas and J. Carillo, vol. 1 (Oxford: Oxford University Press, 2003).

8. K. N. Ochsner et al., 'Rethinking Feelings: An fMRI Study of the Cognitive Regulation of Emotion', *Journal of Cognitive Neuroscience* 14: 1215–29 (2002).

9. D. M. Wegner, R. Erber and S. Zanakos, 'Ironic Processes in the Mental Control of Mood and Mood-Related Thought', *Journal of Personality and Social Psychology* 65: 1093–104 (1993); and D. M. Wegner, A. Broome and S. J. Blumberg, 'Ironic Effects of Trying to Relax Under Stress', *Behaviour Research and Therapy* 35: 11–21 (1997).

10. D. T. Gilbert et al., 'Immune Neglect: A Source of Durability Bias in Affective Forecasting', *Journal of Personality and Social Psychology* 75: 617–38 (1998).

11. Ibid.

12. D. T. Gilbert et al., 'Looking Forward to Looking Backward: The Misprediction of Regret', *Psychological Science* 15: 346–50 (2004).

13. M. Curtiz, *Casablanca*, Warner Bros., 1942.

14. T. Gilovich and V. H. Medvec, 'The Experience of Regret: What When, and Why', *Psychological Review* 102: 379–95 (1995); N. Roese, *If Only: How to Turn Regret into Opportunity* (New York: Random House, 2004); G. Loomes and R. Sugden, 'Regret Theory: An Alternative Theory of Rational Choice Under Uncertainty', *Economic Journal* 92: 805–24 (1982); and D. Bell, 'Regret in Decision Making Under Uncertainty', *Operations Research* 20: 961–81 (1982).

15. I. Ritov and J. Baron, 'Outcome Knowledge, Regret, and Omission Bias', *Organizational Behavior and Human Decision Processes* 64: 119–27 (1995); I. Ritov and J. Baron, 'Probability of Regret: Anticipation of Uncertainty Resolution in Choice: Outcome Knowledge, Regret, and Omission Bias', *Organizational Behavior and Human Decision Processes* 66: 228–36 (1996); and M. Zeelenberg, 'Anticipated Regret, Expected Feedback and Behavioral Decision Making', *Journal of Behavioral Decision Making* 12: 93–106 (1999).

16. M. T. Crawford et al., 'Reactance, Compliance, and Anticipated Regret', *Journal of Experimental Social Psychology* 38: 56–63 (2002).

17. I. Simonson, 'The Influence of Anticipating Regret and Responsibility on Purchase Decisions', *Journal of Consumer Research* 19: 105–18 (1992).

18. V. H. Medvec, S. F. Madey and T. Gilovich, 'When Less Is More: Counterfactual Thinking and Satisfaction Among Olympic Medalists', *Journal of Personality and Social Psychology* 69: 603–10 (1995); and D. Kahneman and A. Tversky, 'Variants of Uncertainty', *Cognition* 11: 143–57 (1982).

19. D. Kahneman and A. Tversky, 'The Psychology of Preferences', *Scientific American* 246: 160–73 (1982).

20. Gilovich and Medvec, 'The Experience of Regret'.

21. T. Gilovich, V. H. Medvec and S. Chen, 'Omission, Commission, and Dissonance Reduction: Overcoming Regret in the Monty Hall Problem', *Personality and Social Psychology Bulletin* 21: 182–90 (1995).

22. H. B. Gerard and G. C. Mathewson, 'The Effects of Severity of Initiation on Liking for a Group: A Replication', *Journal of Experimental Social Psychology* 2: 278–87 (1966).

23. P. G. Zimbardo, 'Control of Pain Motivation by Cognitive Dissonance', *Science* 151: 217–19 (1966).

24. See also E. Aronson and J. Mills, 'The Effect of Severity of Initiation on Liking for a Group', *Journal of Abnormal and Social Psychology* 59: 177–81 (1958); J. L. Freedman, 'Long-Term Behavioral Effects of Cognitive Dissonance', *Journal of Experimental Social Psychology* 1: 145–55 (1965); D. R. Shaffer and C. Hendrick, 'Effects of Actual Effort and Anticipated Effort on Task Enhancement', *Journal of Experimental Social Psychology* 7: 435–47 (1971); H. R. Arkes and C. Blumer, 'The Psychology of Sunk Cost', *Organizational Behavior and Human Decision Processes* 35: 124–40 (1985); and J. T. Jost et al. 'Social Inequality and the Reduction of Ideological Dissonance on Behalf of the System: Evidence of Enhanced System Justification Among the Disadvantaged', *European Journal of Social Psychology* 33: 13–36 (2003).

25. D. T. Gilbert et al., 'The Peculiar Longevity of Things Not So Bad', *Psychological Science* 15: 14–19 (2004).

26. D. Frey et al., 'Re-evaluation of Decision Alternatives Dependent upon the Reversibility of a Decision and the Passage of Time', *European Journal of Social Psychology* 14: 447–50 (1984); and D. Frey, 'Reversible and Irreversible Decisions: Preference for Consonant Information as a Function of Attractiveness of Decision Alternatives', *Personality and Social Psychology Bulletin* 7: 621–26 (1981).

27. S. Wiggins et al., 'The Psychological Consequences of Predictive Testing for Huntington's Disease', *New England Journal of Medicine* 327: 1401–5 (1992).

28. D. T. Gilbert and J. E. J. Ebert, 'Decisions and Revisions: The Affective Forecasting of Changeable Outcomes', *Journal of Personality and Social Psychology* 82: 503–14 (2002).

29. J. W. Brehm, *A Theory of Psychological Reactance* (New York: Academic Press, 1966).

30. R. B. Cialdini, *Influence: Science and Practice* (Glenview, Ill.: Scott, Foresman, 1985).

31. S. S. Iyengar and M. R. Lepper, 'When Choice Is Demotivating: Can One Desire Too Much of a Good Thing?', *Journal of Personality and Social Psychology* 79: 995–1006 (2000); and B. Schwartz, 'Self-Determination: The Tyranny of Freedom', *American Psychologist* 55: 79–88 (2000).

32. J. W. Pennebaker, 'Writing About Emotional Experiences as a Therapeutic Process', *Psychological Science* 8: 162–66 (1997).

33. J. W. Pennebaker, T. J. Mayne and M. E. Francis, 'Linguistic Predictors of Adaptive Bereavement', *Journal of Personality and Social Psychology* 72: 863–71 (1997).

34. T. D. Wilson et al., 'The Pleasures of Uncertainty: Prolonging Positive Moods in Ways People Do Not Anticipate', *Journal of Personality and Social Psychology* 88: 5–21 (2005).

35. B. Fischoff, 'Hindsight =/= foresight: The Effects of Outcome Knowledge on Judgment Under Uncertainty', *Journal of Experimental Psychology: Human Perception and Performance* 1: 288–99 (1975); and C. A. Anderson, M. R. Lepper and L. Ross, 'Perseverance of Social Theories: The Role of Explanation in the Persistence of Discredited Information', *Journal of Personality and Social Psychology* 39: 1037–49 (1980).

36. B. Weiner, "Spontaneous Causal Thinking", *Psychological Bulletin* 97: 74–84 (1985); and R. R. Hassin, J. A. Bargh and J. S. Uleman, 'Spontaneous Causal Inferences', *Journal of Experimental Social Psychology* 38: 515–22 (2002).

37. B. Zeigarnik, 'Das Behalten erledigter und unerledigter Handlungen', *Psychologische Forschung* 9: 1–85 (1927); and G. W. Boguslavsky, 'Interruption and Learning', *Psychological Review* 58: 248–55 (1951).

38. Wilson et al., 'Pleasures of Uncertainty'.

39. J. Keats, letter to Richard Woodhouse, 27 October 1881, in *Selected Poems and Letters by John Keats*, ed. D. Bush (Boston: Houghton Mifflin, 1959).

Chapter 10: Once Bitten

1. D. Wirtz et al., 'What to Do on Spring Break? The Role of Predicted, Online, and Remembered Experience in Future Choice', *Psychological Science* 14: 520–24 (2003); and S. Bluck et al., 'A Tale of Three Functions: The Self-Reported Use of Autobiographical Memory', *Social Cognition* 23: 91–117 (2005).

2. A. Tversky and D. Kahneman, 'Availability: A Heuristic for Judgment Frequency and Probability', *Cognitive Psychology* 5: 207–32 (1973).

3. L. J. Sanna and N. Schwarz, 'Integrating Temporal Biases: The Interplay of Focal Thoughts and Accessibility Experiences', *Psychological Science* 15: 474–81 (2004).

4. R. Brown and J. Kulik, 'Flashbulb Memories', *Cognition* 5: 73–99 (1977); and P. H. Blaney, 'Affect and Memory: A Review', *Psychological Bulletin* 99: 229–46 (1986).

5. D. T. Miller and B. R. Taylor, 'Counterfactual Thought, Regret and Superstition: How to Avoid Kicking Yourself', in *What Might Have Been: The Social*

Psychology of Counterfactual Thinking, ed. N. J. Roese and J. M. Olson (Hillsdale, NJ: Lawrence Erlbaum, 1995), 305–31; and J. Kruger, D. Wirtz and D. T. Miller, 'Counterfactual Thinking and the First Instinct Fallacy', *Journal of Personality and Social Psychology* 88: 725–35 (2005).

6. R. Buehler and C. McFarland, 'Intensity Bias in Affective Forecasting: The Role of Temporal Focus', *Personality and Social Psychology Bulletin* 27: 1480–93 (2001).

7. C. K. Morewedge, D. T. Gilbert and T. D. Wilson, 'The Least Likely of Times: How Memory for Past Events Biases the Prediction of Future Events', *Psychological Science* 16: 626–30 (2005).

8. B. L. Fredrickson and D. Kahneman, 'Duration Neglect in Retrospective Evaluations of Affective Episodes', *Journal of Personality and Social Psychology* 65: 45–55 (1993); and D. Ariely and Z. Carmon, 'Summary Assessment of Experiences: The Whole Is Different from the Sum of Its Parts', in *Time and Decision*, ed. G. Loewenstein, D. Read and R. F. Baumeister (New York: Russell Sage Foundation, 2003), 323–49.

9. W. M. Lepley, 'Retention as a Function of Serial Position', *Psychological Bulletin* 32: 730 (1935); B. B. Murdock, 'The Serial Position Effect of Free Recall', *Journal of Experimental Psychology* 64: 482–88 (1962); and T. L. White and M. Treisman, 'A Comparison of the Encoding of Content and Order in Olfactory Memory and in Memory for Visually Presented Verbal Materials', *British Journal of Psychology* 88: 459–72 (1997).

10. N. H. Anderson, 'Serial Position Curves in Impression Formation', *Journal of Experimental Psychology* 97: 8–12 (1973).

11. D. Kahneman et al., 'When More Pain Is Preferred to Less: Adding a Better Ending', *Psychological Science* 4: 401–5 (1993).

12. J. J. Christensen-Szalanski, 'Discount Functions and the Measurement of Patients' Values: Women's Decisions During Childbirth', *Medical Decision Making* 4: 47–58 (1984).

13. D. Holmberg and J. G. Holmes, 'Reconstruction of Relationship Memories: A Mental Models Approach', in *Autobiographical Memory and the Validity of Retrospective Reports*, ed. N. Schwarz and N. Sudman (New York: Springer-Verlag, 1994), 267–88; and C. McFarland and M. Ross, 'The Relation Between Current Impressions and Memories of Self and Dating Partners', *Personality and Social Psychology Bulletin* 13: 228–38 (1987).

14. William Shakespeare, *King Richard II*, act 2, scene 1 (1594–96; London: Penguin Classics, 1981).

15. D. Kahneman, 'Objective Happiness', in *Well-Being: The Foundations of Hedonic Psychology*, ed. D. Kahneman, E. Diener and N. Schwarz (New York: Russell Sage Foundation, 1999), 3–25.

16. See 'Well-Being and Time', in J. D. Velleman, *The Possibility of Practical Reason* (Oxford: Oxford University Press, 2000).

17. E. Diener, D. Wirtz and S. Oishi, 'End Effects of Rated Quality of Life: The James Dean Effect', *Psychological Science* 12: 124–28 (2001).

18. M. D. Robinson and G. L. Clore, 'Belief and Feeling: Evidence for an Accessibility Model of Emotional Self-Report', *Psychological Bulletin* 128:

934–60 (2002); and L. J. Levine and M. A. Safer, 'Sources of Bias in Memory for Emotions', *Current Directions in Psychological Science* 11: 169–73 (2002).

19. M. D. Robinson and G. L. Clore, 'Episodic and Semantic Knowledge in Emotional Self-Report: Evidence for Two Judgment Processes', *Journal of Personality and Social Psychology* 83: 198–215 (2002).

20. M. D. Robinson, J. T. Johnson and S. A. Shields, 'The Gender Heurristic and the Database: Factors Affecting the Perception of Gender-Related Differences in the Experience and Display of Emotions', *Basic and Applied Social Psychology* 20: 206–19 (1998).

21. C. McFarland, M. Ross and N. DeCourville, 'Women's Theories of Menstruation and Biases in Recall of Menstrual Symptoms', *Journal of Personality and Social Psychology* 57: 522–31 (1981).

22. S. Oishi, 'The Experiencing and Remembering of Well-Being: A Cross-Cultural Analysis', *Personality and Social Psychology Bulletin* 28: 1398–1406 (2002).

23. C. N. Scollon et al., 'Emotions Across Cultures and Methods', *Journal of Cross-Cultural Psychology* 35: 304–26 (2004).

24. M. A. Safer, L. J. Levine and A. L. Drapalski, 'Distortion in Memory for Emotions: The Contributions of Personality and Post-Event Knowledge', *Personality and Social Psychology Bulletin* 28: 1495–1507 (2002); and S. A. Dewhurst and M. A. Marlborough, 'Memory Bias in the Recall of Pre-exam Anxiety: The Influence of Self-Enhancement', *Applied Cognitive Psychology* 17: 695–702 (2003).

25. T. R. Mitchell et al., 'Temporal Adjustments in the Evaluation of Events: The "Rosy View"', *Journal of Experimental Social Psychology* 33: 421–48 (1997).

26. T. D. Wilson et al., 'Preferences as Expectation-Driven Inferences: Effects of Affective Expectations on Affective Experience', *Journal of Personality and Social Psychology* 56: 519–30 (1989).

27. A. A. Stone et al., 'Prospective and Cross-Sectional Mood Reports Offer No Evidence of a "Blue Monday" Phenomenon', *Journal of Personality and Social Psychology* 49: 129–34 (1985).

Chapter 11: Reporting Live from Tomorrow

1. J. Livingston and R. Evans, 'Whatever Will Be, Will Be (Que Sera, Sera)' (1955).

2. W. V. Quine and J. S. Ullian, *The Web of Belief*, 2nd edn. (New York: Random House, 1978), 51.

3. Half of all Americans relocated in the five-year period of 1995–2000, which suggests that the average American relocates about every ten years; B. Berkner and C. S. Faber, *Geographical Mobility, 1995 to 2000*: (Washington, DC: US Bureau of the Census, 2003).

4. The average baby boomer held roughly ten jobs between the ages of eighteen and thirty-six, which suggests that the average American holds at least this many in a lifetime. Bureau of Labor Statistics, *Number of Jobs Held, Labor Market Activity, and Earnings Growth Among Younger Baby Boomers: Results from*

More Than Two Decades of a Longitudinal Survey, Bureau of Labor Statistics news release (Washington, DC: US Department of Labor, 2002).

5. The US Census Bureau projects that in the coming years, 10 per cent of Americans will never marry, 60 per cent will marry just once and 30 per cent will marry at least twice. R. M. Kreider and J. M. Fields, *Number, Timing, and Duration of Marriages and Divorces, 1996* (Washington, DC: US Bureau of the Census, 2002).

6. B. Russell, *The Analysis of Mind* (New York: Macmillan, 1921), 231.

7. The biologist Richard Dawkins refers to these beliefs as *memes*. See R. J. Dawkins, *The Selfish Gene* (Oxford: Oxford University Press, 1976). See also S. Blackmore, *The Meme Machine* (Oxford: Oxford University Press, 2000).

8. D. C. Dennett, *Brainstorms: Philosophical Essays on Mind and Psychology* (Cambridge, Mass.: Bradford/MIT Press, 1981), 18.

9. R. Layard, *Happiness: Lessons from a New Science* (New York: Penguin, 2005); E. Diener and M. E. P. Seligman, 'Beyond Money: Toward an Economy of Well-Being', *Psychological Science in the Public Interest* 5: 1–31 (2004); B. S. Frey and A. Stutzer, *Happiness and Economics: How the Economy and Institutions Affect Human Well-Being* (Princeton, NJ: Princeton University Press, 2002); R. A. Easterlin, 'Income and Happiness: Towards a Unified Theory', *Economic Journal* 111: 465–84 (2001); and D. G. Blanchflower and A. J. Oswald, 'Well-Being over Time in Britain and the USA', *Journal of Public Economics* 88: 1359–86 (2004).

10. The effect of declining marginal utility is slowed when we spend our money on the things to which we are least likely to adapt. See T. Scitovsky, *The Joyless Economy: The Psychology of Human Satisfaction* (Oxford: Oxford University Press, 1976); L. Van Boven and T. Gilovich, 'To Do or to Have? That Is the Question', *Journal of Personality and Social Psychology* 85: 1193–202 (2003); and R. H. Frank, 'How Not to Buy Happiness', *Daedalus: Journal of the American Academy of Arts and Sciences* 133: 69–79 (2004). Not all economists believe in decreasing marginal utility: R. A. Easterlin, 'Diminishing Marginal Utility of Income? Caveat Emptor', *Social Indicators Research* 70: 243–326 (2005).

11. J. D. Graaf et al., *Affluenza: The All-Consuming Epidemic* (New York: Berrett-Koehler, 2002); D. Myers, *The American Paradox: Spiritual Hunger in an Age of Plenty* (New Haven: Yale University Press, 2000); R. H. Frank, *Luxury Fever* (Princeton, NJ: Princeton University Press, 2000); J. B. Schor, *The Overspent American: Why We Want What We Don't Need* (New York: Perennial, 1999); and P. L. Wachtel, *Poverty of Affluence: A Psychological Portrait of the American Way of Life* (New York: Free Press, 1983).

12. Adam Smith, *An Inquiry into the Nature and Causes of the Wealth of Nations* (1776), book 1 (New York: Modern Library, 1994).

13. Adam Smith, *The Theory of Moral Sentiments* (1759; Cambridge: Cambridge University Press, 2002).

14. N. Ashraf, C. Camerer and G. Loewenstein, 'Adam Smith, Behavorial Economist', *Journal of Economic Perspectives* 19: 131–45 (2005).

15. Smith, *The Theory of Moral Sentiments*.

16. Some theorists have argued that societies exhibit a cyclic pattern in which people do come to realize that money doesn't buy happiness but then forget this lesson a generation later. See A. O. Hirschman, *Shifting Involvements: Private Interest and Public Action* (Princeton, NJ: Princeton University Press, 1982).

17. C. Walker, 'Some Variations in Marital Satisfaction', in *Equalities and Inequalities in Family Life*, ed. R. Chester and J. Peel (London: Academic Press, 1977), 127–39.

18. D. Myers, *The Pursuit of Happiness: Discovering the Pathway to Fulfillment, Well-Being, and Enduring Personal Joy* (New York: Avon, 1992), 71.

19. J. A. Feeney, 'Attachment Styles, Communication Patterns and Satisfaction Across the Life Cycle of Marriage', *Personal Relationships* 1: 333–48 (1994).

20. D. Kahneman et al., 'A Survey Method for Characterizing Daily Life Experience: The Day Reconstruction Method', *Science* 306: 1776–80 (2004).

21. T. D. Wilson et al., 'Focalism: A Source of Durability Bias in Affective Forecasting', *Journal of Personality and Social Psychology* 78: 821–36 (2000).

22. R. J. Norwick, D. T. Gilbert and T. D. Wilson, 'Surrogation: An Antidote for Errors in Affective Forecasting' (unpublished manuscript, Harvard University, 2005).

23. Ibid.

24. Ibid.

25. This is also the best way to predict our future behavior. For example, people overestimate the likelihood that they will perform a charitable act but correctly estimate the likelihood that others will do the same. This suggests that if we would base predictions of our own behavior on what we see others do, we'd be dead-on. See N. Epley and D. Dunning, 'Feeling "Holier Than Thou": Are Self-Serving Assessments Produced by Errors in Self- or Social Prediction?', *Journal of Personality and Social Psychology* 79: 861–75 (2000).

26. R. C. Wylie, *The Self-Concept: Theory and Research on Selected Topics*, vol. 2 (Lincoln: University of Nebraska Press, 1979).

27. L. Larwood and W. Whittaker, 'Managerial Myopia: Self-Serving Biases in Organizational Planning', *Journal of Applied Psychology* 62: 194–98 (1977).

28. R. B. Felson, 'Ambiguity and Bias in the Self-Concept', *Social Psychology Quarterly* 44: 64–69.

29. D. Walton and J. Bathurst, 'An Exploration of the Perceptions of the Average Driver's Speed Compared to Perceived Driver Safety and Driving Skill', *Accident Analysis and Prevention* 30: 821–30 (1998).

30. P. Cross, 'Not Can but Will College Teachers Be Improved?', *New Directions for Higher Education* 17: 1–15 (1977).

31. E. Pronin, D. Y. Lin, and L. Ross, 'The Bias Blind Spot: Perceptions of Bias in Self Versus Others', *Personality and Social Psychology Bulletin* 28: 369–81 (2002).

32. J. Kruger, 'Lake Wobegon Be Gone! The "Below-Average Effect" and the Egocentric Nature of Comparative Ability Judgments', *Journal of Personality and Social Psychology* 77: 221–32 (1999).

33. J. T. Johnson et al., 'The "Barnum Effect" Revisited: Cognitive and

Motivational Factors in the Acceptance of Personality Descriptions', *Journal of Personality and Social Psychology* 49: 1378–91 (1985).

34. Kruger, 'Lake Wobegon Be Gone!'

35. E. E. Jones and R. E. Nisbett, 'The Actor and the Observer: Divergent Perceptions of the Causes of Behavior', in *Attribution: Perceiving the Causes of Behavior*, ed. E. E. Jones et al. (Morristown, NJ: General Learning Press, 1972); and R. E. Nisbett and E. Borgida, 'Attribution and the Psychology of Prediction', *Journal of Personality and Social Psychology* 32: 932–43 (1975).

36. D. T. Miller and C. McFarland, 'Pluralistic Ignorance: When Similarity Is Interpreted as Dissimilarity', *Journal of Personality and Social Psychology* 53: 298–305 (1987).

37. D. T. Miller and L. D. Nelson, 'Seeing Approach Motivation in the Avoidance Behavior of Others: Implications for an Understanding of Pluralistic Ignorance', *Journal of Personality and Social Psychology* 83: 1066–75 (2002).

38. C. R. Snyder and H. L. Fromkin, 'Abnormality as a Positive Characteristic: The Development and Validation of a Scale Measuring Need for Uniqueness', *Journal of Abnormal Psychology* 86: 518–27 (1977).

39. M. B. Brewer, 'The Social Self: On Being the Same and Different at the Same Time', *Personality and Social Psychology Bulletin* 17: 475–82 (1991).

40. H. L. Fromkin, 'Effects of Experimentally Aroused Feelings of Undistinctiveness Upon Valuation of Scarce and Novel Experiences', *Journal of Personality and Social Psychology* 16: 521–29 (1970); and H. L. Fromkin, 'Feelings of Interpersonal Undistinctiveness: An Unpleasant Affective State', *Journal of Experimental Research in Personality* 6: 178–85 (1972).

41. R. Karniol, T. Eylon and S. Rish, 'Predicting Your Own and Others' Thoughts and Feelings: More Like a Stranger Than a Friend', *European Journal of Social Psychology* 27: 301–11 (1997); J. T. Johnson, 'The Heart on the Sleeve and the Secret Self: Estimations of Hidden Emotion in Self and Acquaintances', *Journal of Personality* 55: 563–82 (1987); and R. Karniol, 'Egocentrism Versus Protocentrism: The Status of Self in Social Prediction', *Psychological Review* 110: 564–80 (2003).

42. C. L. Barr and R. E. Kleck, 'Self-Other Perception of the Intensity of Facial Expressions of Emotion: Do We Know What We Show?', *Journal of Personality and Social Psychology* 68: 608–18 (1995).

43. R. Karniol and L. Koren, 'How Would You Feel? Children's Inferences Regarding Their Own and Others' Affective Reactions', *Cognitive Development* 2: 271–78 (1987).

44. C. McFarland and D. T. Miller, 'Judgments of Self-Other Similarity: Just Like Other People, Only More So', *Personality and Social Psychology Bulletin* 16: 475–84 (1990).

Afterword

1. Actually, it is a bit unclear just what Bernoulli meant by *utility* because he didn't define it, and the meaning of this concept has been debated for three and a half centuries. Early users of the term were clearly talking about the ability of

commodities to induce positive subjective experiences in those who consumed them. For instance, in 1750 the economist Ferdinando Galiani defined *utilità* as 'the power of a thing to procure us felicity' (F. Galiani, *Della moneta* [On money] [1750]). In 1789 the philosopher Jeremy Bentham defined it as 'that property in any object, whereby it tends to produce benefit, advantage, pleasure, good, or happiness' (J. Bentham, in *An Introduction to the Principles of Morals and Legislation*, ed. J. H. Burns and H. L. A. Hart [1789; Oxford: Oxford University Press, 1996]). Most modern economists have distanced themselves from such definitions – not because they have better ones but because they don't like to talk about subjective experiences. As such, utility has become a hypothetical abstraction of which choices are the measure. If that seems like sleight of mouth to you, then welcome to the club. For more on the history of the concept, see N. Georgescu-Roegen, 'Utility and Value in Economic Thought', *New Dictionary of the History of Ideas*, vol. 4 (New York: Charles Scribner's Sons, 2004), 450–58; and D. Kahneman, P. P. Wakker and R. Sarin, 'Back to Bentham? Explorations of Experienced Utility', *Quarterly Journal of Economics* 112: 375–405 (1997).

2. Most modern economists would disagree with this statement because economics is currently committed to an assumption that psychology abandoned a half-century ago, namely, that a science of human behavior can ignore what people feel and say and rely solely on what people do.

3. D. Bernoulli, 'Exposition of a New Theory on the Measurement of Risk', *Econometrica* 22: 23–36 (1954) (originally published as *'Specimen theoriae novae de mensura sortis'* in *Commentarii Academiae Scientiarum Imperialis Petropolitanae*, vol. 5 [1738], 175–92).

4. Ibid., p. 25.

INDEX

Page numbers in *italics* refer to illustrations.

absences:
    inattention to, 98–104
    perception of, 97–8
absent grief, 152
abstract concepts, mental imagery of, 127–8
actions, inactions regretted more than, 179–80, 191
alexithymia, 62–3
ambiguity:
    of experiences, 159–60, 177
    of stimuli, 154–9
Animal Crackers (comic), *125*
anticipation, 5, 17–18, 19
anxiety, 13–14
appetite:
    prediction of, 115, 117, 136–7, 144, 227
    satiety, 115, 117, 218
    variety, 129–33, *132*, *133*
Aristotle, 34, 36
Arouet, François-Marie (Voltaire), 161
arousal, physiological manifestations of, 58
As You Like It (Shakespeare), 29
attention:
    to absences, 97–104
    focus of, 42–6
auditory cortex, 117, *118*, 121–2
autonomic nervous system, 66
average, self-image distanced from, 229
awareness:
    etymology of, 60
    experience vs., 59–64

Bacon, Francis, 99
Bedazzled (film), 116
Be Here Now (Dass), 16
beliefs:
    transmission of false, 214–22
    about wealth, 217–20
Bentham, Jeremy, 34, 268n
Bentsen, Lloyd, 205
Bergman, Ingrid, 178, 179–80
Bernoulli, Daniel, 235–8, 267n
biases:
    in determination of facts, 163–70, 173
    promoting rationalization, 167–70, 173–4, 259n
    unconscious, 173–4
Bickham, Moreese, 151–2
blindsight, 62, 63
blind spots, 81–2, *82*, 83
Bogart, Humphrey, 178, 179–80
brain function:
    of compensation for missing sensory information, 81–3
    imagination and, 117–20, *118*, *120*, 121–5
    measurement of, 66
    on near future vs. far future, 107
    of object identification, 56–8
    priorities of, 121–5
brain structure:
    anterior cingulate, 63
    emotional awareness in, 63
    evolutionary development of, 10–11, *11*, 15, 56–7
    frontal lobe, 10–16, *11*, 240n

brain structure (*cont.*)
  prefrontal cortex, 13, 66
  prospection capacity vs., 13–16
  sensory experience vs. awareness in,
    62
Bush, George H. W., 205, 206
Bush, George W., 209–10, 230

California:
  assumptions on well-being in,
    103
  political affiliations in, 206
Candide (Voltaire), 161
capital punishment, 169
Casablanca (film), 178, 179–80
Cather, Willa, vii
causation:
  absences considered in analysis of,
    99
  emotional influence of knowledge
    of, 185–91
changes, sensitivity to, 138
children:
  in comprehension of future, 9,
    239n
  realism in, 85–6
  as source of parental happiness,
    220–2, 221
Cicero, Marcus Tullius, 36–7,
    99–100
Clarke, Arthur C., 112
Clarke's first law, 112
Clever Hans (horse), 172–3
Clinton, Bill, 206
coaching, learning through, 195–6,
    211, 233
Coleman, Ornette, 49
colour:
  memory of, 40–1, 42–3
  vision, 31–2
communication:
  of inaccurate beliefs, 216–17
  as vicarious observation, 213
comparisons, 137–46
  with the past, 136–41
  with the possible, 140, 141–3

presentism biases in, 143–6
selective, 166–7
conjoined twins, 29–30, 46–50, 53
consciousness:
  in animals, 4, 5–6, 60
  as emergent property, 67
  of future, 4–6
control:
  innate desire for, 20–3
  prospection motivated by urge for,
    20–4
Coolidge, Calvin, 253n
corrigibility:
  defined, 193
  of hindsight, 191–2
  personal experience as source of,
    196–211
  secondhand knowledge as basis of,
    212–33
Cymbeline (Shakespeare), 111, 195

Dawkins, Richard, 265n
Day, Doris, 212
decisions:
  Bernoulli's formula for, 235–8,
    267n–8n
  in economic comparisons, 137–43,
    145
  emotion vs. logic in, 121
  irrevocability vs., 184–5
  regret as factor in, 178–80
declining marginal utility, 130, 218,
    265n
defence mechanisms, 162–3, 257n
  see also immune system,
    psychological; positive
    attitudes
delays, psychological pain of, 107
Dennett, Daniel, 59
depression, 13, 22, 124, 152
Descartes, René, 63–4
differentiation, 229–32
disabilities, happiness vs., 104, 152,
    153
distant future, imagination of, 105–8,
    249n

Index

see also future; imagination;
  prediction; prospection
Dukakis, Michael, 205–6
Durant, Will, 85

Eastman, George, 76, 77, 92, 93–4
eating:
  imagination of, 115, 117, 119–20,
    144, 227
  satiety limit on, 218
  variety in, 129–33, *132*, *133*
economic status, cultural beliefs on
    happiness vs., 217–20, 236–7,
    265*n*, 266*n*
Egas Moniz, António, 12–13
Einstein, Albert, 172
elections:
  of 1988, 205–6
  of 1992, 113–14
  of 2000, 209–10, *209*
emergent properties, 67–8
emotional happiness, 31–5, 38–9
  see also happiness
emotions:
  biased memories of, 114–15,
    207–10
  in decision process, 121
  gender differences vs., 207–8
  imagination of, 119–25, *120*
  lack of awareness of, 62–3
  measurable physiologic expressions
    of, 65–6
  misidentification of, 55, 58–63
  negative overestimates in prediction
    of, 175–80
  uniqueness attributed to, 232
empiricism, 163
endings, disproportionate emphasis
    on, 202–5
Epicurus, 36
Essay on Man (Pope), 33
eudaimonia, 36
evolutionary biology:
  on gene transmission, 215–16
  of human brain, 10–11, *11*, 15,
    56–7

experience:
  ambiguity of, 159–60, 177
  anticipation of, 17
  awareness vs., 59–64
  Cartesian view of, 63–4
  causation of, 185–91
  circumstances associated with,
    185–6
  education vs., 196
  ending overemphasized in memory
    of, 202–5
  etymology of, 60
  frequent vs. rare, 197–201
  inescapable, 183–5
  learning from, 196–7, 211
  pleasure diminished in repetition of,
    129–33, *132*, *133*
experience-stretching hypothesis,
    50–3, *51*, 69

Farnsworth, Philo T., 49
filling in:
  of imagination, 90–4, 95, 108, 114,
    238
  in memory, 79–81, 113–14
  of perception, 81–3
  present knowledge used in, 113
Fischer, Adolph, 75, 76, 77, 92–3, 94
Ford, Gerald, 206
frequency:
  of remembered experience,
    197–201
  stimulus ambiguity vs., 157
Freud, Anna, 257*n*
Freud, Sigmund, 34, 146, 257*n*
frontal lobe function, 10–16, *11*,
    240*n*
frontal lobotomy, 13
future:
  children's understanding of, 9, 239*n*
  human uniqueness in concepts of,
    4–6, 9
  near vs. distant, 105–8
  percentage of thoughts about, 16
  underestimation of novelty of,
    111–13

future (*cont.*)
  see also imagination; prediction; prospection
futurist literature, 111–12

Gage, Phineas, 11–12, *12*, 240n
gain, assessments of loss vs., 145–6
Galiani, Ferdinando, 268n
genes, transmission of, 215–16
goodness, emotions as criteria of, 71
Gore, Al, 206, 209–10
Greene, Graham, 58

habituation, 130–3, 253n
Hamlet (Shakespeare), 151
happiness:
  of Californians, 103
  child rearing as source of, 220–2, 221
  conscious efforts toward, 174
  current experience vs. memory of, 40–2
  defined, 31–8
  economic status vs., 217–20, 236–7, 265n, 266n
  as emotional state, 31–5, 38–9
  ethnic background vs., 208
  explanations vs., 186–91
  false cultural beliefs on sources of, 217–22, 221
  judgmental meaning of, 37–8
  machine, 35, 244n
  measuring, 65–6
  moral attitudes toward, 35–7, 38–9
  predictions of, 209–10, 209
  as subjective state, 31–2, 39, 46–8, 52–4
  universal desire for, 33–5
health:
  control as beneficial to, 21–2
  immune function in, 162
  therapeutic writing exercises and, 186–7
  see also illness

hearing:
  brain function and, 117–18, 121–2
  compensation for absent sounds in, 82–3
Henry VI, Part III (Shakespeare), 235
Hitchcock, Alfred, 212
Hobbes, Thomas, 34
Holmes, Sherlock (fictional character), 96–7, 108
horse, intelligence attributed to, 172–3
  see also Clever Hans
Horton, Willie, 205
hunger, prediction of, 115, 117, 123–4, 136–7
Hussein, Saddam, 248n

idealism, 85–9
illness:
  cancer, 18, 167
  chronic, 104, 153
  death vs., 153
imagination:
  brain function and, 117–20, *118*, *120*, 121–5
  defined, 77–8
  of distant future, 105–8, 249n
  ease of, 24, 89–90
  of emotional responses, 119–25, *120*
  factors not considered by, 101–4, 108, 225
  mental priorities vs., 121–5
  of pleasant scenarios, 17–18
  presentism as influence on, 24, 114–17, 123–6, 134–7, 143–6, 147, 226–7
  sensory, 117–20, *118*, *120*, 252n
  of sequential experiences, 129–33, *133*
  shortcomings of, 24–5, 76–8, 223, 224–8, 238
  spatial metaphors of, 128–9, 133, 134
  specific details added through, 90–4, 95, 108, 238

surrogation vs., 223–6, 226, 266n
as unique human ability, 5
visual, 122–3, 127–8, 134
see also prediction; prospection
immune system, psychological, 162,
227
on actions vs. inactions, 179–80
inescapable circumstances as trigger
of, 183–5
intensity as trigger of, 180–3
positive attitudes as, 160–70
unawareness of, 173–4, 175,
177–8, 179, 182, 191
inactions, actions regretted more than,
179–80, 191
individuality, 224, 229–32, 233
inescapability, 183–5
information, biased processing of,
164–70, 191

Jefferson, Thomas, 146
judgmental happiness, 37–8
Julius Caesar (Shakespeare), 3, 96

Kant, Immanuel, 85, 94
Keats, John, 190–1
Kelvin, William Thomson, 112
Kerry, John, 206, 230
King Lear (Shakespeare), xiii
knowledge:
false beliefs transmitted as,
214–22
firsthand vs. secondhand, 195–6,
213–14

language, positivity of words in, 33
language-squishing hypothesis, 46–7,
47, 51–2, 69
large numbers, law of, 67–70
'Lavandou, Le' (Cather), vii
Lennon, John, 89
Locke, John, 84–5
loss, assessments of gain vs., 145–6

Man Who Knew Too Much, The
(film), 212
marriage:
anticipation of, 105
parenthood satisfaction vs., 221,
221
rates of, 214, 265n
Marx, Harpo, 78, 79
Marx Brothers, 78
mastery, development of, 195–6
meaning, 155–7
Measure for Measure (Shakespeare),
55
measurement, 64–71
memes, 265n
memory:
in brain function, 117–18, *118,
120, 122*
of colour, 40–1, 42–3
of common occurrences vs. unusual
experiences, 197–201
compression process of, 78–9, 197,
202
corrigibility through, 191–2
current experience vs., 40–2,
113–14
endings overemphasized in, 202–5
faulty reconstruction of, 41–2,
79–81, 197, 206–10
of pain, 113, 202–4
presentism as distortion factor in,
113–14
theoretical influences on, 207–10
menstrual cycle, emotional influences
of, 207, 208
mental imagery:
atemporal aspect of, 134
of concrete objects vs. abstract
concepts, 127–8
vision vs., 122–3
see also imagination
metaphors, 128–9
Midsummer Night's Dream, A
(Shakespeare), 75
Mill, John Stuart, 34, 35
Miller, George, 88

mistakes, repetition of, 196–7, 201

morality, happiness vs., 35–7, 38–9

moral philosophy, 71

Müller-Lyer lines, xv, *xv*

Necker cube, xv, *xv*, 157–8, *158*, 159, 160

negative events, resilient responses to, 151–3, 227–8

Newcomb, Simon, 112

nexting, 6–9

Nixon, Richard, 206

N.N. (head-injury patient), 14–15, 16

Nozick, Robert, 35, 244*n*

objects, identification of, 56–8

optical illusions, xv–xvi, *xv*, 23–4, 157–8, *158*

optimistic expectations, 18, 161–3

see also positive attitudes

Osten, Wilhelm von, 172–3

ownership, losses vs. gains of, 145–6

pain:

anticipation of, 19

circumstances associated with, 185–6

ending experiences overemphasized in recollections of, 202–4

hypnotic suggestion of, 59

intensity of, 180–3

memory of, 113, 202–4

in near future vs. distant future, 107

see also pleasure

Pangloss (fictional character), 161

parenthood, personal satisfactions of, 220–2, 221

Pascal, Blaise, 34–5

perception:

of absence, 97–8

failures of, 42–6, 86

filling-in process in, 81–3

initial realist interpretations of, 86, 87–8

preexisting knowledge combined with, 85–6

Perot, H. Ross, 113–14

Pfungst, Oskar, 172–3

physical disabilities, 104, 152, 153

Piaget, Jean, 85–6, 88

Plato, 34, 36, 71, 249*n*

pleasure:

anticipation of, 17–18

habituation to, 129–33, *132*, 253*n*

satiety vs., 218

see also happiness; pain

political opinions, 168, 169, 206

politics, electoral, 113–14, 205–6, 209–10, 209

Pope, Alexander, 33

positive attitudes, 160–70

analytic bias utilized for, 167–70

of cancer patients, 18, 167

credibility vs., 161–3

on inactions vs. actions, 179, 191

toward inescapable circumstances, 183–5

as psychological defence, 162–3

selective information in support of, 163–7, 191

suffering intensity vs., 180–3

unconscious perpetration of, 173–4

Pound, Ezra, 49

practice, learning through, 195–6, 211, 233

predictions:

additional assumptions as influences in, 91–2

current conditions as biasing factor in, 111–13, 123–6, 144, 147, 226–7

of desire, 115, 116–17, 136–7, 144, 227

details ignored in, 102–3

of happiness, 209–10, 209
memory shortcomings vs., 200–1
negative emotions overestimated in, 175–80
of next event, 5–9
present feelings as starting point in, 134–7
of reactions to current events tragedies, 248n–9n
of reactions to rejection, 175–8, 176
of regrets, 178–80
surrogates' current experiences as basis of, 223–8, 226, 232, 233, 266n
see also future; prospection
prefeeling, 120–5, 120
prefrontal cortex, 13, 66
presentism:
comparison evaluations distorted by, 143–6
defined, 109
emotional evaluation affected by, 123
in historical assessments, 146
imagination influenced by, 114–17, 123–6, 134–7, 143–6, 147, 226–7
memory biased by, 113–14
price:
comparisons of, 139–40
value vs., 236–7
probability theory, 68
proof, standards of, 168–70, 173
prospection:
brain structure and, 13–16
children's development of, 9, 239n
as control mechanism, 20–4
defined, 1
emotional aspects of, 17–19
illusions of, 23–4
next-event prediction vs., 5–9
as unique human trait, 4, 15–16
see also future; prediction
psychological defences, 162–3, 257n

see also immune system, psychological; positive attitudes
psychological experiments, ethical guidelines on, 241n–2n

Quayle, Dan, 205

rare events:
emotional impact of, 188
memory of, 197–201
rationalization:
ambiguity exploitation and, 154–60
defined, 149
in resilient responses to trauma, 151–3
surrogation vs., 227–8
see also immune system, psychological; positive attitudes
reading, 6–7, 59–60, 61
Reagan, Ronald, 206
realism:
defined, 73, 88
distorted illusions vs., 161–3, 171
idealism vs., 85–9
sensory perceptions as conduit of, 84–5
reality-first policy, 121–5
recency, of stimulus, 157
Reeve, Christopher, 151, 152
regret, prediction of, 178–80
rejection, prediction of reactions to, 175–8, 176
relativism, economic, 137–43
repetition:
of mistakes, 196–7, 201
pleasure diminished with, 129–33, 132
resilience, 152–3
risk, of loss vs. gain, 145–6
Russell, Bertrand, 214

satiety, 115, 117, 218

Schappel, Lori, 29–30, 46–7, 47, 48, 49–50, 51, 53

Schappel, Reba, 29–30, 46–7, 47, 48, 49–50, 51, 53

Schindler's List (film), 201–2

science:
    analytical favouritism in, 259n
    erroneous predictions in, 112
    measurement in, 64, 65–6
    observational bias in, 163, 164
    psychology as, 64
    random sampling in, 164

senses:
    brain function and, 117–20, *118, 120*
    realistic information conveyed through, 84–5

sequential experiences, imagination of, 129–33, *132, 133*

Shackleton, Ernest, 53–4

Shakespeare, William, xiii, 3, 29, 55, 75, 96, 111, 127, 151, 172, 195, 203, 212, 235

Shrader, K. P., 259n

Smith, Adam, 218, 219

social stability, cultural beliefs in service of, 233
    economic imperatives as, 217, 219–20
    parenthood satisfactions as, 222

Socrates, 35, 36

Solon, 36, 48

sounds, brain compensation for, 82–3

space, time conceptualized with metaphors of, 128–9, 133, 134

Spielberg, Steven, 202

stimulus:
    meaning vs. ambiguity of, 155–9
    objective vs. subjective, 154–6

Stravinsky, Igor, 174

subjective states, 29–72
    awareness deficiencies on, 58–63

current experiences vs. memories of, 40–2
    of emotional happiness, 24, 31–2, 39, 46–8, 52–4
    irreducibility of, 32, 243n
    measurement of, 64–71
    recent experience as bias in evaluation of, 48–53

subjectivity:
    defined, 27
    of interpretation of stimuli, 155–9
    realistic reassessments of, 88

super-replicators, 215–16, 217, 222

surrogation:
    as effective prediction aid, 223–8, 226, 232, 233, 266n
    rejection of, 224, 228–32

temporal distance, 105–8

thirst, prediction of, 123–4

time:
    as abstract concept, 127–8
    mental images dissociated from, 134
    repetition over, 129–33, *132, 133*
    spatial metaphors used for, 128–9, 133, 134

Tolstoy, Lev, 33

traumatic events:
    resilient reactions to, 151–3
    writing therapy after, 186–7

Troilus and Cressida (Shakespeare), 172, 212

twins, conjoined, 29–30, 46–50, 53

uncertainty, 188–91

uniqueness, overestimation of, 224, 229–32, 233

utility, 235–7, 267n–8n

value, price vs., 236–7

van Gogh, Vincent, 49

variety, pleasure affected by, 129,
130–3, *132*, *133*, 252n–3n
Venus and Adonis (Shakespeare), 127
virtue, happiness vs., 35–7
vision:
    attention focus in, 42–4
    brain function and, 62, 117,
    118–19, *118*, *120*, 121–2
    of colour, 31–2
    compensation for blind spot in,
    81–2, *82*, *83*
    of distant objects, 104–5, 106–7,
    249n
    experience dissociated from
    awareness in, 62
    mental imagery vs., 122–3, 252n
    object identification process in,
    56–7
visual cortex, 117, *118*, 121–2

visual discontinuities, 43
Voltaire (François-Marie Arouet), 161

wealth, happiness vs., 217–20, 236–7,
    265n, 266n
'Whatever Will Be, Will Be (Que Sera,
    Sera)' (Livingston and Evans), 212
women:
    emotionality attributed to, 207–8
    historical sexist views of, 146
    maternal satisfaction levels of, 221
Wonderful Wizard of Oz, The
    (Baum), 83–4
Wright, Jim, 151
Wright, Wilbur, 112

Zappa, Frank, 32

PERMISSIONS ACKNOWLEDGMENTS

Excerpt from the song lyric 'Whatever Will Be, Will Be (Que Sera, Sera),' words and music by Jay Livingston and Ray Evans. Copyright © 1955 by Jay Livingston Music, Inc. and St. Angelo Music. Copyright renewed. International copyright secured. All rights reserved. Reprinted by permission of Jay Livingston Music, Inc. and Universal Music Publishing Group.

ILLUSTRATION CREDITS

Figures 1, 2, 6, 7, 8, 9, 10, 11, 13, 14, 15, 16, 17, 18, 19, 21, 22, 23, and 24 created by Mapping Specialists, Ltd.

Figure 3 reprinted from J. M. Harlow, 'Recovery from the Passage of an Iron Bar Through the Head,' *Publications of the Massachusetts Medical Society* 2: 327–47 (1868).

Figure 12: *Animal Crackers* cartoon by Fred Wagner. Copyright © 1983 by Tribune Media Services. Reprinted by permission of Tribune Media Services.

Figure 14 reprinted from C. Walker, 'Some Variations in Marital Satisfaction,' in R. Chester and J. Peel, eds., *Equalities and Inequalities in Family Life* (London: Academic Press, 1977), 127–39. Copyright © 1977. Reprinted by permission of Elsevier.

Figure 19 adapted with permission from the American Psychological Association from D. T. Gilbert, E. C. Pinel, T. D. Wilson, S. J. Blumberg, and T. P. Wheatley, 'Immune Neglect: A Source of Durability Bias in Affective Forecasting,' *Journal of Personality and Social Psychology* 75: 617–38 (1998).

Figure 20 reprinted with permission from the American Psychological Association from T. D. Wilson, D. B. Centerbar, D. A. Kermer, and D. T. Gilbert, 'The Pleasures of Uncertainty: Prolonging Positive Moods in Ways People Do Not Anticipate,' *Journal of Personality and Social Psychology* 88: 5–21 (2005).

Figure 22 adapted with permission from Guilford Publications, Inc. from T. D. Wilson, J. Meyers, and D. T. Gilbert, 'How Happy Was I, Anyway? A Retrospective Impact Bias,' *Social Cognition* 21: 407–32 (2003).

P.S.

Ideas,
interviews
& features...

About the author

2 Q and A with Professor Gilbert

5 Life at a Glance

10 A Writing Life

11 Top Ten Favourite Electric Guitarists

About the book

12 Confessions of a Spoondigger by Professor Daniel Gilbert

Read on

16 If You Loved This, You Might Like...

17 Find Out More

Q and A with Professor Gilbert

What was your childhood like?
My dad was a professor (still is, in fact) and my mother was a writer and artist, so I got the best of both worlds. Like my mother, I write. Like my father, I am a scientist. My childhood was lovely because my parents were wonderful, but also because my older brother was the first child and my younger sister was the first girl, and because I wasn't the first anything I received far less scrutiny, which meant that I could do as I pleased generally because no one was looking.

As a child, what did you want to be when you grew up?
I seem to recall that second baseman for the Chicago Cubs and *Playboy* photographer were both very high on the list. At the age of 10, I thought these were the two best ways to ensure that I'd eventually get to see a woman naked.

Would you describe yourself as happy – and where on the eight-point rating?
People are terrible at remembering how happy they were, predicting how happy they'll be, or judging how happy they are in general. They can, however, tell you how happy they are at the moment they are asked, so at this moment I am somewhere around 6.5. Am I usually this way? I don't know. People tell me I have a sunny disposition but that I am easily annoyed.

What makes you happy?
This is another question that people have trouble answering for themselves. When you are truly happy you aren't noticing how happy you are, which makes it difficult to recall later. With that said, I have the sense that I am happiest when I am writing without interruption. I can wake up at 5 a.m., walk directly to my desk, and write for 10 hours without ever remembering to eat or brush my teeth. My normal day is an endless series of interruptions – email, telephone, family, students – so the occasional uninterrupted day is (I think) my greatest pleasure.

What makes you unhappy?
I get snippy and sarcastic when people use language incorrectly. I shouldn't, but I do. When a clerk at a store says, 'That will be three dollars', I say, 'Really, when?' I know, I know. I should be shot.

You say that research shows that having children doesn't make us happier. Do you think becoming a dad made you happier?
Intuitions and data often collide. The data say the earth is round, but it looks flat to me. My intuition is that fatherhood increases my average daily happiness, but the data say that unless I'm different from most people, this probably isn't so. Of course, I'm not exactly like other people inasmuch as I am 48 and my son is 30, so perhaps I'm free to believe my intuitions – in which case, I ▶

6 My book isn't meant to make people happy. It is meant to make them smart about happiness by telling them what science has discovered. I'm not in the business of telling people what's right. I'm in the business of helping them see what's true and then letting them decide for themselves what to do about it. 9

3

Q and A *(continued)*

◄ believe that he makes me happy and that my granddaughter makes me even happier.

Can self-help books help?
Yes. They help the authors make money. Anyone who takes psychological advice from someone who isn't trained and licensed to give it should have their head examined. Would anyone buy a book written by a cab driver called *How to Remove Your Own Appendix*?

Does what you know about the way the human brain works in any way help you to be happy?
Knowing that people overestimate the impact of almost every life event makes me a bit braver and a bit more relaxed because I know that whatever I'm worrying about now probably won't matter as much as I think it will.

Do you intend for your book to help people to think differently?
My book isn't meant to make people happy. It is meant to make them smart about happiness by telling them what science has discovered. I hope to give people information that they can use (or not use) as they wish. I'm not in the business of telling people what's right. I'm in the business of helping them see what's true and then letting them decide for themselves what to do about it.

What do you hope to accomplish with your research?
I'd like to say that I am trying to understand

❛ Self-help books help their authors make money. Anyone who takes psychological advice from someone who isn't trained and licensed to give it should have their head examined. ❜

4

errors in affective forecasting so that we can learn how best to overcome them. The trouble is that forecasting errors are not clearly a 'disease' that requires a 'cure'. Indeed, some people have suggested that inaccurate forecasts may play an important role in our lives. Having said that, I'm willing to bet that on balance we are best served by accurate estimates of the emotional consequences of pains, tragedies and embarrassments. However, at heart I'm just a guy who is curious about human nature, and what I really want from my research is a deeper understanding of who we are and what we are doing here. If my research has a practical benefit, I'm happy about that. If it doesn't, I'm not even slightly worried. What is the practical benefit of knowing how the universe began, or of understanding the evolution of the mealworm?

Have you tried self-help books, meditation, therapy, whatever, to do battle with your brain's behaviour?
Yes, I call it single malt scotch therapy and I find that at the end of a long day it tames my brain quite nicely.

Would you like to live in the eternal now?
No. I enjoy remembering the past and imagining the future. My ability to do these things is among nature's greatest gifts to me, so why would I want to be rid of it? Anyone who wants to live in the moment should have been born a mosquito. ▶

Author photograph © Marilynn Oliphant

LIFE *at a Glance*

BORN
1957, Ithaca, New York, USA

EDUCATED
BA in Psychology at the University of Colorado in Denver (1981); PhD in Psychology at Princeton University in New Jersey (1985)

CAREER
Professor at the University of Texas in Austin, 1985–96; Professor at Harvard University, 1996–present

FAMILY
Married to Marilynn Oliphant. Has a son, Arlo Gilbert, and a granddaughter, Daylyn Gilbert.

LIVES
Cambridge, Massachusetts, with his wife and a lack of pets.

Q and A *(continued)*

◄ **You say we all imagine our futures will be better than our presents, but Americans particularly so. Why do you think Americans are more optimistic than other nationalities?**

America is a young country that has had an astonishingly rapid rise to the height of wealth and power. Things have been good here since the beginning and have always gotten better, so Americans believe that this is their eternal trajectory. I see a few surprises in store for us.

Why do you open every chapter of your book with a Shakespeare quote?

There are two reasons. First, throughout history there have been wonderfully insightful people who have made shrewd guesses about how the mind works. Modern science allows us to decide which of these guesses was right and which was wrong. Shakespeare has a pretty good track record of being right, so I decided to let him kick off each chapter. The second reason is that I am an ordinary low-brow guy who prefers action movies to sonnets and Tater Tots to pâté, but in my daily life I impersonate a Harvard professor, and I always have this nagging feeling that to play this role properly requires that I be a refined snob who sits around reading Shakespeare while eating little sandwiches without crusts. That's not me, but the quotes will fool everyone – provided you don't rat me out.

6 Knowing that people overestimate the impact of almost every life event makes me a bit braver and a bit more relaxed because I know that whatever I'm worrying about now probably won't matter as much as I think it will. 9

What is your favourite film or book and why?

I like movies a lot, but I love old science fiction movies, so let me say *The Day the Earth Stood Still*. I still get shivers when someone says 'Klaatu, barada nikto'. As far as books go, I'm very fond of T. C. Boyle. *Drop City* was terrific. And A. M. Homes. She's dark and strange in all the right ways. *Music for Torching* made me feel like I was an alien on my own planet.

Are you an optimist?

No, sorry, I don't know the first thing about making eyeglasses.

We all suffer from illusions of foresight. How do you imagine your future? Do you think you are wrong about it?

When I think about the future I imagine quitting my job and sneaking off to a Hawaiian island with my wife and my guitar and never coming back. I suppose that makes me an optometrist after all.

You say that we regret not doing something more than something we did. What do you regret not doing – and doing?

I regret not looking after my health a bit better back when it was easy to do. The guy who had my body before me wasn't all that nice to it. I don't have any Great Regrets of Action, though I suppose I would take back every instance in which I made someone I love feel bad if that were possible. ▶

❝ Three good things I do regularly: (1) drink freshly ground double-strength coffee, (2) walk to and from my office every day, and (3) listen to Miles or Jimi at least once a week. ❞

Q and A *(continued)*

◄ **You say it's the frequency not the intensity of positive events in your life which makes you happy. What positive events reoccur in your life regularly and therefore contribute to your happiness?**
Three good things I do regularly: (1) drink freshly ground double-strength coffee made in a French press every morning, (2) walk to and from my office every day, and (3) listen to Miles or Jimi at least once a week (and if you have to ask their last names then you are even less cool than I am).

Do you think that too much choice in modern life is making us miserable?
Well, we surely have too many stupid choices. A year or two ago I bought a dozen pairs of identical cargo pants and identical black T-shirts and now when I wake up in the morning I never think about what to wear. Why should we waste our lives deciding whether to have Coke or Pepsi, with or without caffeine, with or without sugar, with or without lemon, in a can or a bottle or a litre or a cup, with or without ice, and a straw thank you?

Do you think we have lost some primal ignorance that would have kept us happy?
No, no, no. Did I mention no? Every generation has the illusion that things were easier and better in a simpler past. Dead wrong. Things are better today than at any time in human history. Our primal ignorance is what keeps us whacking each other over the head with sticks, and not what allows us to paint a Mona Lisa or design a

❛ Modern life comes with too many stupid choices. A year or two ago I bought a dozen pairs of identical cargo pants and identical black T-shirts and now when I wake up in the morning I never think about what to wear. ❜

space shuttle. The 'primal ignorance that keeps us happy' gives rise to obesity and global warming, not antibiotics or the Magna Carta. If human kind flourishes rather than flounders over the next thousand years, it will be because we fully embraced learning and reason, and not because we surrendered to some fantasy about returning to a world that never really was.

What do you consider to be your greatest achievement and why?
I am rather proud of being listed just before Dizzy Gillespie on the list of the Most Famous High School Dropouts. I mean, just to be on the same page as Diz...wow.

What would you do right now if you learned you were going to die in 10 minutes?
I'd go to the phone to say goodbye to a few important people. Then, if I had time, I'd smoke up a storm. ■

A Writing Life

When do you write?
Mornings – 5 a.m. to noon is prime time.
I can't stand to talk to other people in the
morning but I like to talk to myself, and
that's what writing is.

Why do you write?
That's like 'Why do you breathe?'
Answer: 'How could I not?'

Pen or computer?
Computer. Pens are for fast writers or slow
thinkers and I'm neither.

Silence or music?
Silence, instrumental music (I have to be
able to hear the narrator in my head
and singers compete with him) or crowd
noise (the background hum of people
and machinery at a café is perfect).

How do you start a book?
With a detailed plan that I abandon as soon
as the book develops its own trajectory.

And finish?
When they pry it out of my cold, dead
hands. I could keep fussing with a piece of
writing forever. Someone (e.g. my editor)
has to make me stop.

Do you have any writing rituals?
Orient desk 45 degrees from full moon and
consume powdered wombat intestines.
Actually, no.

❛ I can't stand
to talk to other
people in the
morning but I
like to talk to
myself, and
that's what
writing is. ❜

What or who inspires you?
I write about science because I am inspired by the rational conquest of myth, prejudice and ignorance.

If you weren't a writer, what would you do?
Weep?

What's your guilty reading pleasure? Favourite trashy read?
I never feel guilty about reading. Watching *Star Trek* on television...now that's guilt. ■

TOP TEN FAVOURITE ELECTRIC GUITARISTS

1   *John Abercrombie*

2   *Nels Cline*

3   *Bill Frisell*

4   *Jimi Hendrix*

5   *Charlie Hunter*

6   *John McLaughlin*

7   *Terje Rypdal*

8   *Sonny Sharrock*

9   *Ralph Towner*

10   *Stevie Ray Vaughan*

Confessions of a Spoondigger

by Professor Daniel Gilbert

Some writers hate writing but love having written. I'm the opposite. Writing *Stumbling on Happiness* was the most enjoyable project I've ever done, and now that it's over, my life is a little bit empty. I never thought I would feel this way.

When I sat down at a Starbucks in New York City one day in 1998 and wrote the first page of this book, I didn't know what I was doing or why. I just wrote something I was thinking about, and then wrote something else, and by the time I'd finished the first chapter... well, I still didn't have any idea what book it might be the first chapter of. In fact, the more I wrote, the more convinced I became that there wasn't a book to write – not because I wasn't up to the task, but because the raw material for a book on prospection simply wasn't there. The theories I wanted to talk about hadn't yet been formulated, and the data I wanted to describe hadn't yet been collected. So I stuck the chapter in a drawer and wondered if I'd ever get back to it.

Two years later, things had changed. Scientists of every kind had become interested in happiness and prospection, and they were producing ideas and experiments at a remarkable rate, providing me with the material I needed but didn't have (or at least didn't know about) two years earlier. So I picked up that old first chapter, rewrote it and began to write more.

I never stopped again. I'm a terribly slow writer – I agonize over word order and word choices and word sounds and word plays, and I will often rewrite a single paragraph for hours – but the writing wasn't the hard part. The hard part was developing a way to organize hundreds of ideas and findings from various fields and weave them together into a coherent narrative that could get from the first page to the last page under its own steam. I didn't want to write a series of essays. I wanted to tell a single story that had twists and turns and cul-de-sacs, but that also had a clear narrative arc. In his review, Malcolm Gladwell called my book 'a psychological detective story' and I was gratified by that description because it suggests that the book has a visible plot, and that's what I'd struggled to accomplish.

I struggled to accomplish something else as well. Psychology is the most fascinating topic one could possibly write about, and yet psychologists generally manage to make it as dull as paste. I didn't want to write in that cold, detached, phony, imperious 'Dr Science' voice that we all use when we write our research papers. I needed a different voice, and when I looked around the only one I could find was my own. So I used it. One of my colleagues told me the other day that his wife read my book and said: 'This is just like having a conversation with Dan except that you don't get to talk back' ▶

> ❛ Doing science is a bit like flying in an airplane over a vast landscape for an hour, and then landing the plane, getting out and digging in the dirt with a teaspoon for twenty years. ❜

Confessions . . . *(continued)*

◄ (to which my colleague responded: 'And how is that different from any conversation with Dan?'). She's right. I talk exactly like the guy in the book. Admirers of my book call it personal, warm and funny, and critics call it juvenile, self-indulgent and annoying. I suspect that all these adjectives describe me pretty well.

I worked on this book for five years, rarely showing a word of what I'd written to anyone. I wish I could say that I'd done it for you, but I didn't. I did it for me. When I was a graduate student, I studied with a dear and brilliant man named Ned Jones. Ned and I would have long, rambling conversations about research, and after a few hours he would invariably stop and say, 'So, Danny boy, what's the big picture?' And then he'd answer his own question by articulating a simple, beautiful core truth around which our entire conversation had been a set of baroque variations. Although I inherited Ned's love for the big picture, the everyday work of a scientist doesn't leave much time for it. Doing science is a bit like flying in an airplane over a vast landscape for an hour, and then landing the plane, getting out and digging in the dirt with a teaspoon for twenty years. Establishing facts is so damned time-consuming that as the years pass it is easy to forget you have wings. I'd been spoon-digging for 25 years, and I wrote this book because I wanted to take a plane ride and see what my life had been about. I'd published dozens of studies that were vaguely related to each other, I'd read thousands of studies that were vaguely

> 6 I didn't want to write in that cold, detached, phony, imperious "Dr Science" voice that we all use when we write our research papers. I needed a different voice, and when I looked around the only one I could find was my own. So I used it. 9

related to mine, but I didn't really know the shape of the landscape on which each was a small hill. *Stumbling on Happiness* is a report from me to me – a way of telling myself what I've been doing all these years.

Now I know and you do too, so it's time for me to polish up my spoon and get back to the dirt. I enjoyed telling you this story. Thanks for listening. ∎

6 In his review, Malcolm Gladwell called my book "a psychological detective story" and I was gratified by that description because it suggests that the book has a visible plot, and that's what I'd struggled to accomplish. 9

If You Loved This,
You Might Like...

Strangers to Ourselves
by Timothy D. Wilson
The psychologist who collaborates with
Professor Gilbert in his research examines
the limits of what we know – and can know
– about ourselves.

The Tipping Point or Blink
by Malcolm Gladwell
The world's best social-science journalist
writes thoughtfully and engagingly about
psychology.

Freakonomics
by Steven D. Levitt and Stephen J. Dubner
A maverick economist and an award-
winning journalist team up to show how
economic analysis can shed light on a variety
of interesting and unusual problems.

The Happiness Hypothesis
by Jonathan Haidt
With warmth and insight, a psychologist
describes the intersection of scientific
psychology and ancient wisdom.

Happiness: A History
by Darrin M. McMahon
In this intellectual tour de force, a historian
examines how the concept of happiness has
evolved over two thousand years.

Find Out More
Website

At **danielgilbert.com** you will find Professor
Gilbert's research articles, which form the
skeleton of his book.